SCREENING EQUIPMENT HANDBOOK

SCREENING EQUIPMENT HANDBOOK

for industrial and municipal water and wastewater treatment

SECOND EDITION

Tom M. Pankratz

CRC Press
Taylor & Francis Group
Boca Raton London New York

CRC Press is an imprint of the
Taylor & Francis Group, an **informa** business

CRC Press
Taylor & Francis Group
6000 Broken Sound Parkway NW, Suite 300
Boca Raton, FL 33487-2742

© 1995 by Taylor & Francis Group, LLC
CRC Press is an imprint of Taylor & Francis Group, an Informa business

First issued in paperback 2019

No claim to original U.S. Government works

ISBN-13: 978-0-367-44921-6 (pbk)
ISBN-13: 978-1-56676-256-4 (hbk)

Visit the Taylor & Francis Web site at
http://www.taylorandfrancis.com

and the CRC Press Web site at
http://www.crcpress.com

Library of Congress Cataloging-in-Publication Data

Main entry under title:
Screening Equipment Handbook: For Industrial and Municipal Water and Wastewater
      Treatment—Second Edition

Library of Congress Card Number 94-62045

For Michael H. Pankratz

Table of Contents

Foreword

WHEN it was first published in 1988, *Screening Equipment Handbook* fulfilled a real need: at last, engineers and operators had a useful reference that identified the large variety of screens available for water and wastewater applications. Screening seemed simpler then, and most screens had rather narrow, clearly defined roles.

Screens could be neatly divided into easy-to-categorize groups. Most in-channel screens had coarse openings. Most fine screens required pumping and resulted in higher than desired headlosses. Some wastewater treatment plants were still designed without any screens at all! The uses and misuses of screens seemed reasonably clear, and *Screening Equipment Handbook* reflected this.

The last six years have brought many changes to the screening equipment industry. The screens that were once precisely and simply categorized are now available in a bewildering variety of designs, sizes, and types. As a result of ingenuity, an increased interest in pollution prevention, the beneficial use of biosolids, and increasingly stringent legislation, the once-clear delineations that included opening size, design, and usage classifications have become blurred.

More consultants are designing municipal wastewater plants in innovative ways, such as substituting fine screens for primary clarifiers. Screens known for their use in pulp and paper mills now regularly appear in wastewater plants attached to vegetable processing facilities.

It's easy to become confused when considering which type of screen might be best for a particular application. There's so much available in the marketplace because the predictable childlike screening market has become an adventurous adolescent. This revision of the Handbook is timely and reflects the many changes that the industry has undergone.

When you have a new project to consider, use this updated book as your reference. Keep your eye firmly on your project's goal and draw on the experience of consulting engineers, respected manufacturers, and their representatives. You'll find there's a screen to help you accomplish your goal. From protecting downstream equipment to capturing by-products for resale, from designing a new grass-roots plant to shoehorning a new screen into an old plant to expand capacity, today's screen market offers the solution you need, and this book will help you identify your options.

ROBIN REDDY
Lake Bluff, Illinois

Preface

THE first edition of *Screening Equipment Handbook* was published in 1988 and was the first comprehensive reference to review the water and wastewater screening equipment industry. The industry has since undergone many changes. Although these changes may not reflect startling new technological advances, they do include improvements in overall screening efficiency, reliability, and economics.

The objective of this second edition of *Screening Equipment Handbook* is to summarize the available screening equipment options and provide a consolidated source of basic design and application information to assist the engineer in selecting a screen best suited for the particular application.

Although every chapter has been completely updated, significant changes have also been made to the chapters dealing with trash rakes and fine screens, and new chapters have been added to introduce screenings handling and comminuters/grinders. Many new photographs and diagrams have been added to illustrate equipment designs and product features, and the number of entries in the Manufacturers' Directory has been almost doubled to include more than 115 up-to-date names and addresses of equipment manufacturers.

The book is divided into chapters that describe specific types of screening or screening-related products. Chapter 1 reviews common screening equipment applications and identifies the types of screens most frequently employed in each. Chapters 2, 3, 4, and 5 review trash rakes, traveling water screens, drum screens, and passive screens, respectively. These screens are most frequently associated with the screening of raw intake water. Bar screens and fine screens are covered in Chapters 6 and 7, which emphasize the use of screens in preliminary and primary wastewater treatment applications. Chapter 8 introduces screenings handling products. Although some of these products have been available for several years, their use continues to increase and represents one of the most significant changes in wastewater screening. Chapter 9 describes comminuters and grinders and their use in the water and wastewater industry, while Chapter 10 reviews microscreen design and application. The appendices include a variety of information that I have found helpful over the years.

After completing the first edition of the book, I continued to find new information that hadn't been included. I frequently wished that publication had been delayed for one or two months, or more. When I mentioned this to my son Chad, he wisely directed me to the following passage from Hemingway's *Death in the Afternoon:*

> I was not able to write anything about the bullfights for five years – and I wish I would have waited for ten. However, if I had waited long enough I probably never would have written anything at all since there is a tendency when you really begin to learn something about a thing not to want to write about it but rather to keep on learning about it always, and at no time, unless you are very egotistical, which of course accounts for many books, will you be able to say: now I know all about this and will write about it. Certainly I do not say that now; every year I know there is more to learn, but I know some things which may be interesting now, and I may be away from bullfights for a long time and I might as well write what I know about them now.

As new information becomes available, there will be contemplation of a new, third edition.

I would like to thank Robin Reddy and Mike Toepfer for reviewing this new edition and providing many helpful suggestions and comments.

In addition to the people and companies who provided illustrative materials for the first edition of this book, I would like to thank the following individuals for their help with this edition: Lars Apelqvist, Tom Baber, R. Baillorgeon, Peter Blake, John Booth, Pascal Bovagnet, Tom Brown, Tim Canfield, Wayne Cassell, Dean Chang, Richard Coniglio, Lee Cook, Larry Crowell, Roger Davidson, Randy Delenikos, Theo de Wolff, J. M. Douglas, Jeff Drake, Mike Drake, B. R. Evans, Erich Fink, Dieter Frankenberger, Rich Gargan, Dennis Geran, P. A. Goeman, Brian Graham, Stacey Grimes, Gary Haggard, H. B. Haffer, Malcolm Haigh, A. Hanson, Dennis Herold, Henry Hunt, P. E. Jackson, William Jungman, Wolfgang Krahn, James Lageman, Ernest Latal, Bill Lauritch, Don Losacco, Gary Leudtke, Vernon Lucy, Gary Mackey, Robert Manwaring, W. David Martin, Werner Marzluf, Jean-Jacques Maurel, G. Meneghetti, Gabriel Meunier, John Mullin, Yusuf Mussalli, A. Neuhold, Delmar Nichols, Finn Nielsen, A. Nilsson, Philip Orrill, Ken Ohyler, Bill Palarz, Becca Pankratz, Joseph Pastore, Alfred Patzig, Tom Quimby, Robin Reddy, L. Reichenau, Stanley Rudzinski, Karl Heinz Rusch, Mr. Schaaf, Kristy Schloss, James Siler, Nico Smits, Rich Sommers, Michael Spring,

C. L. Sprinkle, Judy Stevenson, Fred Tipton, C. H. Van Leeuwen, Mark Watson, Thomas Wingfield, H. Wirth and Jake Zelenietz.

Finally, I would like to thank my wife Julie, and our children Chad, Sarah, Mike, and Katie for their continued support and encouragement.

TOM PANKRATZ
League City, Texas

CHAPTER 1

Screen Applications

T HE application of the best screening system for a particular project is somewhat of an art. Although there are many "classic" applications, where one type of screen is obviously better suited than any other, the selection process for many projects is far more complicated.

Screen opening size and flow rate are the most important criteria used in the selection of a screening system. Other categories include civil, equipment, and operational costs; plant hydraulics; debris handling requirements; operator qualifications; and availability.

The following categories describe common screening applications within particular industries and the types of screens that they most frequently employ. It is often possible to successfully utilize screens other than the types mentioned below.

ELECTRIC POWER GENERATION

Nuclear and fossil fueled steam electric generating plants require approximately 500 to 1500 gallons of water per minute (32 to 96 L/s) per megawatt of rated capacity. This water is primarily used for cooling purposes in the plant's condensers and must be screened prior to use to prevent clogging of tubes and openings in the surface and jet condensers, as well as interfering with the proper operation of the circulating and condenser pumps.

Many power plants discharge the condenser exhaust directly to a lake, ocean, or river that serves as a heat sink. Most newer power plants recycle their condenser exhaust through cooling towers. Approximately 3% to 5% of the recirculating water must be continually replaced to make up for losses due to evaporation, drift, and blowdown. Such a "closed cycle" cool-

Power plant intake screens, courtesy of Screening Systems International.

ing water system greatly reduces intake water requirements and, consequently, the number of screens.

Traveling water screens and drum screens are used to fulfill the screening requirements at most power plants that use a surface cooling water source. A large power plant that utilizes this "once through" approach in its cooling water system may require as many as forty-eight individual traveling water screens. These screens are usually preceded by a coarse trash rack, equipped with a trash rake mechanism, to remove very large and heavy debris.

Environmental conditions may require screens to be fitted with special features to minimize their adverse affect on fish and other marine life. Many traveling water screen manufacturers offer "fish screen" modifications for these applications.

Some power plants have successfully applied passive screening systems for surface water intakes. These screens virtually eliminate problems associated with debris handling/removal and may make compliance with environmental regulations easier.

Hydroelectric power stations in the northwestern United States may use traveling water screens to screen the pumped recirculating water on a fish ladder system. This is a relatively low flow application compared to the other previously described power plant screen uses and usually requires two or three screens.

Static screens and manually cleaned bar screens have been used successfully to dewater spray water produced by traveling water screens and drum screens.

WOOD PRODUCTS INDUSTRY

Pulp and paper plants are intensive users of raw water, requiring up to 20,000 gallons (76 m³) of water per ton of product. Many pulp, paper, and saw mills generate their own electricity and use traveling water screens to screen raw condenser feedwater and process water prior to its use.

Most mills process the logs used as raw materials for their product. The logs are conveyed from the woodyard to large debarking drums via a flume. Prior to being recycled, the flume water passes through a grit chamber followed by one or more traveling water screens, static screens, or rotary fine screens to remove bark, leaves, and twigs that have fallen off or been removed during handling.

Some paper plants and kraft mills utilize static screens and rotary fine screens in their waste treatment process, prior to clarification, to remove and/or recover floating and suspended fibers and other particles that might hinder sedimentation. Other fine screen applications in paper and pulp mills include broke thickening, save-all, and whitewater screening.

MISCELLANEOUS MANUFACTURING AND PROCESSING

Food processing, steel mills, petro-chemical plants, refineries, textile, and other large manufacturing plants also require the use of mechanical

Poultry processing, courtesy of Hycor Corp.

screening equipment. Requirements for many of these applications are very site-specific.

Fine screens are also commonly used for solids' separation or product recovery. Selection of the proper type of fine screening equipment is highly dependent on the volume and consistency of the solids to be screened.

Traveling water screens are usually used to screen fire water, condenser and process cooling water, boiler feedwater, potable water, and process water. For lower flow application, mechanically cleaned bar screens have also been used successfully.

IRRIGATION PROJECTS

Irrigation and other open-channel water distribution projects in the western and southwestern United States may use traveling water screens, belt screens, or bar screens to protect pumping equipment at remote pump stations.

WATER TREATMENT PLANTS

Large surface water treatment plants and desalination plants may employ traveling water screens, drum screens, bar screens, or passive screens

Irrigation pump station, courtesy of FPI, Inc.

Wastewater treatment plant, courtesy of Infilco Degremont.

to screen raw intake water as the first step in the treatment process. Inadequate screening may result in damage to downstream equipment, may increase chemical requirements, and may hinder the treatment process.

WASTEWATER TREATMENT PLANTS

One or more bar screen and/or fine screen can be found at virtually every wastewater treatment plant in the world. The type of screens used ranges from manually cleaned, coarse bar screens to fully automated microscreens.

Screens may be used in preliminary, primary, and tertiary treatment processes in a wastewater treatment plant. The use of fine screens for the screening of sludge, grease, scum, and solids produced in other treatment processes has increased significantly since the mid-1980s.

Application examples and suggestions are further discussed in the Bar Screen, Fine Screen, Drum Screen, and Microscreen chapters of this book.

COMBINED SEWER OVERFLOW

The use of screens in Combined Sewer Overflows (CSO) applications are increasing since publication of the U. S. EPA's draft policy requiring

Stormwater overflow, courtesy of Longwood Engineering.

CSO's to install nine minimum controls, including the control of solid and floatable materials in CSO discharges. One of the most effective methods of meeting this requirement is to install screening equipment at the CSO discharge. Bar screens, fine screens, drum screens, and traveling water screens have all been used successfully in CSO systems.

CHAPTER 2

Trash Rakes

TRASH rakes are heavy duty raking devices that are used to remove large or rough debris retained on a trash rack. They protect pumping equipment at electric generating plants, desalination plants, pump stations, and flood control projects and may be used as a preliminary screening device to protect finer screens such as traveling water screens or drum screens.

A trash rake consists of one or more stationary trash rakes and a screen raking mechanism. Bar spacings on the trash rack usually range from 1-1/2″ to 4″ (38 to 100 mm). Most racks are constructed of carbon steel bars, but at least one manufacturer offers trash racks constructed entirely of high density polyethylene polymers. This rack weighs up to 75% less than its carbon steel counterpart, and is less prone to problems resulting from frazil ice, marine growths, and corrosion.

Trash racks can be equipped with sensors to detect the presence of scuba divers or boats in the intake area.

There are a variety of raking mechanisms available for use in a variety of intake configurations, including installation on vertical building and dam walls. Rakes may be mounted on fixed structures designed to clean a single trash rack, suspended from an overhead gantry, or wheel-mounted to traverse the entire width of an intake structure and clean individual sections of a very wide trash rack.

Trash rakes have been designed with effective rake widths of up to 20 ft (6 m). The largest cleaning depth reported is 350 ft (164 m).

Some mechanically cleaned bar screens, including the reciprocating rake and catenary types, have been adapted for use in trash rake applications.

Because of the large variations in bar spacings, debris loading, and the application of trash rakes, there is no established relationship for the quantity of screenings removed per unit volume of flow.

7

Trash rake installation, courtesy of Bieri Hydraulik.

Trash screen installation, courtesy of Infilco Degremont.

CABLE OPERATED TRASH RAKES

This trash rake design uses a single, toothed rake that is raised and lowered by means of one or more sets of stainless steel cables to clean a stationary trash rack. The rack is lowered down the upstream face of the trash rack with the teeth in the open, disengaged position. As the rake reaches the channel bottom, the teeth are rotated to their closed position, meshing with the stationary bars. The rake is then lifted up the face of the trash rake, collecting accumulated debris. As the rake reaches the discharge position above the operating deck elevation, the teeth are opened, discharging debris into a trough or hopper.

RAKE HOIST

The upper ends of the 3/8″ (9 mm) minimum diameter rake cables are fixed to a shaft-mounted hoist drum. The drum's line shaft operates in self-

Cable operated trash rake, courtesy of Hans Kunz GesmbH.

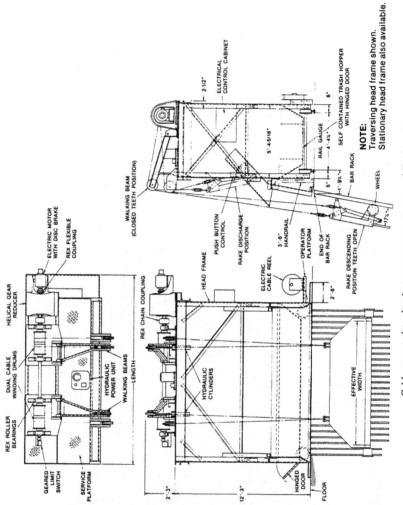

ELECTRIC MOTOR WITH DISC BRAKE

REX FLEXIBLE COUPLING

HELICAL GEAR REDUCER

DUAL CABLE WINDING DRUMS

REX ROLLER BEARINGS

GEARED LIMIT SWITCH

SERVICE PLATFORM

HYDRAULIC POWER UNIT

WALKING BEAMS

LENGTH

REX CHAIN COUPLING

HEAD FRAME

PUSH BUTTON CONTROL

RAKE DISCHARGE POSITION

ELECTRIC CABLE REEL

3'-6" HANDRAIL

OPERATOR PLATFORM

END OF BAR RACK

RAKE DESCENDING POSITION TEETH OPEN

HYDRAULIC CYLINDERS

EFFECTIVE WIDTH

HINGED DOOR

FLOOR

2'-3"

12'-3"

2'-6"

WALKING BEAM (CLOSED TEETH POSITION)

2-1/2"

ELECTRICAL CONTROL CABINET

5'-4-5/16"

RAIL GAUGE

4'-4¼"

6'-1"

6'

1'-9¾"

17¼"

SELF CONTAINED TRASH HOPPER WITH HINGED DOOR

BAR RACK

WHEEL

NOTE:
Traversing head frame shown.
Stationary head frame also available.

Cable operated trash rake arrangement, courtesy of Envirex, Inc.

Wall-mounted trash rake, courtesy of Bieri Hydraulik.

Trash rake with grab crane, courtesy of Hans Kunz GesmbH.

aligning roller bearings and is directly coupled to a drive unit furnished with a motor mounted brake. Rotation of the hoist drum changes the effective length of the cables and results in the raising or lowering of the rake. The excess cable that accumulates during the rake's ascent is stored on the periphery of the grooved hoist drum. Limit switches automatically stop the rake at the top and bottom of its travel. Hoist speeds may range from 15 to 40 ft (4.5 to 12 m) per minute. Two speed hoist drives (slow descent, fast ascent) may be used for deep trash rake installations to reduce rake cycle times.

Standard duty trash rakes are furnished with a single hoist drum and one set of cables to lift objects up to 17 " (430 mm) in diameter, with hoist capacities to 2000 lb (900 kg). Heavy duty units may be capable of lifting objects up to 24 " (600 mm) in diameter and are furnished with dual hoist drums and two sets of cables, with hoist capacities to 4600 lb (2000 kg). The hoist mechanism should be equipped with a slack cable safety switch and an over-travel switch. Hoist drive motors typically range from 1 to 7.5 horsepower (0.75 to 5.6 kW).

RAKE TOOTH OPERATION

The cleaning cycle begins when the rake is lowered to the bottom of the channel and the rake teeth are rotated to the closed, engaged position. Most designs allow the rake teeth to penetrate the bar rack a minimum of 1/2 " (13 mm).

Tooth opening and closing are accomplished by changing the effective cable length. This may be done by controlling the slippage between a fixed and free segment of the hoist drums. This system requires that the inside drum, or drum segment, be keyed to the drum shaft. The other drum is free for limited rotation controlled by the fixed drum. The free drums are equipped with band brakes to allow for the opening of the rake.

An alternative arrangement for operating the rake teeth uses a walking beam assembly. This system requires a small hydraulic power unit to operate a hydraulic cylinder that is used to raise and lower a walking beam assembly, changing the effective length of the tooth control cables. Hydraulic relief valves automatically open the rake teeth if an overload condition occurs.

GUIDED RAKE

Each end of the rake assembly may be equipped with wheels or sliding blocks to operate in guide channels embedded in the walls of the concrete channel. The guided rake design is recommended for vertical trash racks or trash racks inclined less than ten degrees from vertical.

NONGUIDED RAKE

The nonguided rake design allows the rake to rise over obstructions on the trash rake and is preferred for most trash rake installations. This design requires the rake assembly to be equipped with flanged wheels, each wide enough to ride on two bars. Nonguided rake designs are recommended to be used with trash rack inclinations of ten degrees or more.

TRAVERSING MECHANISM

Trash rakes can be mounted on a self-propelled hoist frame capable of traversing the width of an entire intake. This allows the trash rake to be positioned to clean a specific portion of a wide trash rack.

A drive unit can propel the screen hoist frame at a speed of approximately 25 ft (7.6 m) per minute on flanged wheels riding on tee-rails. Tie-down clamping brackets are used to engage the downstream rail and prevent the unit from tipping as large debris loads are lifted. The hoist frame is usually equipped with an operator's platform or cab and a push-button control station from which the unit is operated. A pendant control station may also be furnished.

The rake's transition from the trash rack to its discharge position on the hoist frame is accomplished through the use of a deadplate constructed as an integral part of the hoist framework. For most installations, the deadplate should be designed to operate over a 3'-6" (1 m) high handrail. A spring-loaded electric cable reel with sufficient cable to span the intake structure must be provided to store the electric power cable.

DEBRIS TROUGH OR HOPPER

Debris collected by the rake can be discharged into a trough, debris hopper, or conveyor located on the downstream side of the trash rack. Debris hoppers are often fabricated as an integral part of the hoist frame and should be furnished with a hinged bottom or side door for debris clean-out. The debris hopper may be wheel- or rail-mounted and designed so that it can be removed from the frame entirely. Commercial "dumpsters" may also be utilized as trash rake hoppers.

Some trash rake machines are equipped with a hydraulic grab crane to unload debris from the unit's hopper. The grab crane may also be used to remove large floating debris in front of the trash rack.

CHAIN-OPERATED TRASH RAKE

Chain-operated trash rakes use toothed rakes or debris lifters mounted

Chain-operated trash rake, courtesy of Duperon.

on endless strands of chain, operating over head and foot sprockets. The rakes clean the trash rack as they ascend, and discharge debris on the top, downstream side of the rack prior to beginning their descent. Debris is usually gravity discharged, but rake wipers or strippers may be used to help remove tenacious debris.

Chain-operated trash rakes are available in front and back cleaned designs. Front cleaned designs, where both the ascending and descending chains are located on the upstream side of the trash rack, should employ some means to prevent debris from jamming between the chain and lower sprockets. This may include the use of a chain guide and enclosed sprocket housing or the use of a "catenary" instead of a lower sprocket. (See the description of the Catenary Bar Screen in Chapter 6 of this book.) Chain-operated trash rakes are available with electric and hydraulic drive motors.

HYDRAULIC TRASH RAKE

A hydraulically operated rake mechanism may be used to clean trash accumulated on the face of a trash rack. There are several variations of hydraulic trash rakes available with lifting capacities of up to 2500 lb (1130 kg), for use with trash racks up to 65 ft (20 m) deep.

Chain-operated trash rake, courtesy of Hydropower Turbine Systems.

Chain-operated trash rake, courtesy of Hydropower Turbine Systems.

These units consist of a single rake that can be raised and lowered and opened and closed by means of two or more hydraulic cylinders. Hydraulic trash rakes can be furnished as stationary units or can be motor driven to traverse the width of an intake.

HINGED-ARM

The operation of this trash rake resembles that of a hydraulically operated excavator or backhoe. The synchronized motion of the hydraulic cylinders is used to extend the rake out in front of the trash rack, lower it to the channel bottom, and lift debris out of the water for disposal at the deck elevation. The operator may control the movement of the pivoting-arm rake mechanism with a pendant-type hand controller or stationary push-button station or while sitting in a fully enclosed cab equipped with a joystick-type controller. Some designs allow the upper unit to rotate 360 degrees so that debris can be lifted out of the channel and unloaded in any direction.

Although they may be stationary, most of the rake units are mounted on tee-rails to be electrically driven along the width of the intake. They may also be furnished with rubber tires and a diesel or gasoline powered engine that allows them to be driven to other locations. Other available options include climate-controlled cabs and various grab or clamshell rake bucket configurations.

Maximum rake widths vary with specific screen designs, ranging from

Hinged-arm trash rake, courtesy of Smalley Excavators, Inc.

Hinged-arm trash rake, courtesy of Landustrie Sneek BV.

17

4 to 12 ft (1.2 to 3.7 m). Units can operate in water depths of up to 23 ft (7 m). Electric drive motor sizes range from 5 to 30 horsepower (3.7 to 24.4 kW). Diesel-operated units may range to 45 horsepower (33.5 kW).

TELESCOPING ARM

Another trash rake design utilizes a rake attached to the bottom of a carriage-mounted telescoping arm. The arm may be raised and lowered by means of a chain lift mechanism or a hydraulic cylinder. Engagement and disengagement of the rake are accomplished by a hydraulic cylinder that tilts the rake carriage into, or away from, the trash rack. The carriage may be mounted on a horizontal rail to transverse the width of the intake.

OVERHEAD GRAB RAKE

An overhead grab-type trash rake utilizes an elevated track to support a grab rake above the trash rack. As the grab rake is lowered, its tines engage the trash rake and clean the rack from top to bottom. The rake

Telescoping arm rake, courtesy of Atlas Polar.

Chain-operated trash rake, courtesy of Brackett Green.

hydraulically closes when it reaches the channel bottom, and the lift motor is activated. Debris removed from the channel can be deposited directly into a container or truck.

The grab rake is carriage-mounted so that it can be traversed the length of the track and be positioned over the portion of the trash rack that requires cleaning. The elevated track can be curved to suit most intake configurations and transport debris to any convenient location.

Rake widths range up to 8 ft (2.4 m) and have lifting capacities of up to 1100 lb (500 kg). An alternate design has the grab rake suspended from a swing-arm mounted on a central column.

ROTARY BRUSH

Another trash rack cleaning device utilizes a submersible rotary brush to remove marine growths, including zebra mussels, barnacles, and algae accumulations. The 16″ to 36″ (400 to 900 mm) diameter brush uses synthetic or nonferrous metal bristles and ranges in length from 3 to 12 ft (1 to 3.7 m). The brush is raised and lowered on the face of the trash rack by means of a gantry crane or other hoisting mechanism. A shaft-mounted drive motor operates the brush at 60 to 200 rpm.

CHAPTER 3

Traveling Water Screens

TRAVELING water screens, or bandscreens, are automatically cleaned screening devices that are used to remove floating or suspended debris from a channel of flowing water. Traveling water screens are used to protect pumping or other downstream equipment from objectionable debris in surface water intakes and other applications.

Traveling water screens consist of a continuous series of wire mesh panels bolted to basket frames, or trays, and attached to two matched strands of roller chain. The chain operates in a vertical path over head- and footsprockets, carrying the baskets down into the water, around the footsprockets, and back up through the water, over the headsprockets. As raw water passes through the revolving baskets, debris is collected and retained on the upstream face of the wire mesh panels. The larger particles of debris are collected on a 2″ to 3″ (50 to 76 mm) wide lifting shelf that forms the lower, or trailing, edge of the basket frame.

The debris-laden baskets are lifted out of the flow and above the operating floor where a high-pressure water spray is directed outward through the mesh to remove impinged debris. The spray wash water and debris are collected in a trough for further disposal.

The path of the chain and basket assemblies is guided within a track that also forms a skeleton-like frame to support the screen structure. A labyrinth seal between the screen frame structure and channel wall prevents debris from passing around the screen. Each basket is furnished with two end seal plates to seal the openings between the ends of the basket and the chain guide tracks.

Traveling water screens are usually not operated until a specified differential headloss or time lapse has occurred. Some applications may be operated more frequently, even continuously, based on specific design or application requirements.

There are three distinct types of traveling water screens available: the "Thru Flow," "Dual Flow," and "Center Flow." In the thru flow design, the

21

Traveling water screen, courtesy of FMC Corp.

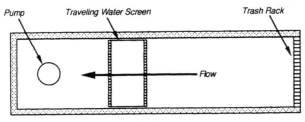

Thru Flow Screen

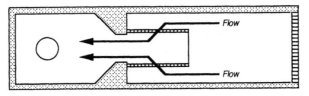

Dual Flow Screen

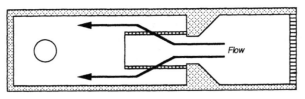

Center Flow Screen

Traveling water screen flow arrangements.

screening surface is oriented perpendicular to the flow, with only the ascending baskets utilized as available screening area. Dual flow designs orient the screening surface parallel to the flow to utilize both the ascending and descending baskets as active screening area. The center flow is a variation of the dual flow design, where water enters the center of the screen structure and exits outwards through the ascending and descending baskets.

Traveling water screens can be manufactured to suit virtually any raw water screening application. Screens are available with basket widths that range from 2 to 14 ft (0.6 to 4.3 m) and channel depths up to 100 ft (30 m) or more. Wire mesh openings have historically ranged from 1/4" to 3/4" (6 to 19 mm), but some recent applications have required openings as small as 1/16" (2 mm).

THRU FLOW TRAVELING WATER SCREENS

Thru flow traveling water screens, also referred to as "direct," "single," or "uni" flow screens, are the most common type of raw water intake screen in the United States. Over 12,500 thru flow traveling water screens have been manufactured since 1895.

The terminology used to describe this traveling water screen refers to the water's flow pattern through it. A thru flow screen is installed in a channel with the screening surfaces oriented perpendicular to the water flow. Raw water passes through the ascending and descending screen baskets, respectively, utilizing the ascending baskets on the upstream (front) side of the screen as active screening surfaces. As debris is elevated out of the flow, it is removed by a high-pressure water spray before the cleaned baskets begin their descent on the downstream (rear) side of the screen.

Debris is prevented from passing around the screen through the use of vertical guides that also serve to locate the screen in the channel. Sealing plates are mounted on both ends of each basket to prevent solids from passing around the sides of the baskets as they rotate around the footsprocket and ascend in the chain guide tracks.

Thru flow screens typically utilize 80 to 90% of the channel width as effective screening area and occupy less than 5'-6" (1.7 m) of channel length in direction of the flow. The straight-through flow pattern usually means that no elaborate forebay is required and that any required pumps can be located relatively close to the screens.

The primary advantage of the thru flow screen (versus dual flow designs) is the reduced cost of intake construction and equipment. Intakes that incorporate thru flow screen designs tend to be compact, with simple and uncomplicated flow patterns.

The potential for debris carryover is the most important disadvantage of the thru flow traveling water screen. If the spray system does not effectively remove debris from the ascending baskets, the debris will be carried over into the flow on the downstream side of the screen. This defeats the purposes of the screen and may result in damage to downstream equipment. The only protection from carryover problems is proper design, maintenance, and monitoring of the screen and its spray wash system. Because only one-half of the total submerged screening area (the ascending baskets) is utilized, a thru flow screen installation will require more/larger screens to pass a given flow than a dual flow installation. This generally results in higher operating and maintenance costs.

DUAL FLOW TRAVELING WATER SCREENS

Dual flow traveling water screens were developed in the 1920s and have been used extensively in Europe and Asia and are being applied on an in-

Thru flow traveling water screen, courtesy of Screening Systems International.

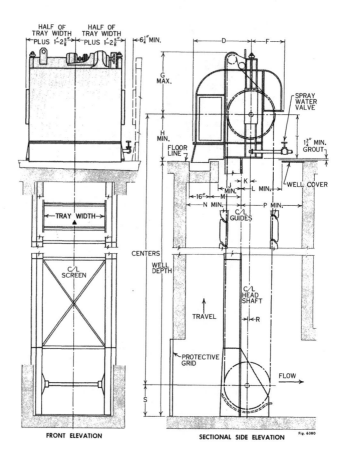

Model screen	D	F	G	H	J	K	L	M	N	P	R	S	T
					FEET AND INCHES								
45A	4-4⅛	2-9¾	4-11½	3-5½	1-8¾	0-11¾	3-7½	2-2¾	3-9	5-0	0-1½	2-5½	3-3
46A	3-8⅝	2-2⅜	4-1½	3-0	1-6¼	0-6⅞	2-7¼	2-0¼	3-6½	3-11¾	0-1¼	2-0	2-9

Thru flow traveling water screen, courtesy of FMC Corp.

creasingly frequent basis in the United States. Dual flow and thru flow traveling water screens differ in the water's flow pattern through the screen. The screening surfaces of dual flow screens are oriented parallel to the water flow, with raw water passing through both the ascending and descending screen baskets. Screened water exits through an opening in the center of the screen structure.

A dual flow traveling water screen is mechanically similar to a thru flow screen that has been rotated ninety degrees in the channel.

This flow pattern offers some significant advantages over thru flow designs. One advantage is that the entire submerged screen surface is utilized as active screen area. This means that a dual flow screen of a given width will pass almost twice as much water, at the same velocity, as a thru flow screen of the same basket width.

Even more important is the fact that the dual flow designs virtually eliminate the possibility of debris carryover. The debris remains on the upstream side of the screening surface on both the ascending and descending baskets at all times. The boot section design of dual flow screens also eliminates boot sealing problems associated with thru flow designs.

The primary disadvantage of the dual flow traveling water screen is the increased construction costs due to the larger concrete channel and support structure. The screen orientation and the flow pattern of water will

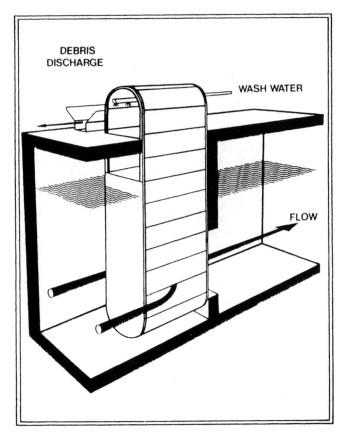

Dual flow screen arrangement, courtesy of Brackett Green.

Dual flow traveling water screen, courtesy of Hubert Stavoren BV.

also require careful hydraulic consideration to prevent high localized ve-locities resulting from abrupt changes in the direction of the water flow.

Applications where severe debris loading conditions are anticipated should also consider that debris collected on the descending basket run will block the screen for the remainder of the operating cycle. This may re-

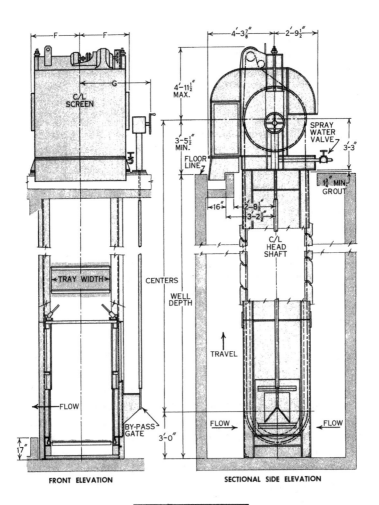

Tray width	F	G
FEET AND INCHES		
3-0	2- 8⅝	4- 1⅛
3-6	2-11⅝	4- 4⅛
4-0	3- 2⅝	4- 7⅛
4-6	3- 5⅝	4-10⅛
5-0	3- 8⅝	5- 1⅛
5-6	3-11⅝	5- 4⅛
6-0	4- 2⅝	5- 7⅛
6-6	4- 5⅝	5-10⅛
7-0	4- 8⅝	6- 1⅛
7-6	4-11⅝	6- 4⅛
8-0	5- 2⅝	6- 7⅛

Dual flow traveling water screen, courtesy of FMC Corp.

quire continuous operation and/or structural modification to insure that the screen will withstand higher and more frequent differential headloss conditions.

Although the industry standard basket width for thru flow screens is 10 ft (3 m), most dual flow screens have a basket width of 5 to 7 ft (1.5 to 2.1 m). The narrower basket widths result in an overall reduction in the weight of the baskets and chain and, consequently, a reduction in sprocket, chain, and bearing wear. The shorter unsupported span of the structural members in a narrower basket also makes them less vulnerable to damage from a headloss.

Modifications can be made to the sealing arrangements of some dual flow designs that will allow the use of mesh openings as small as 0.04″ (1 mm).

It is generally accepted that the initial cost of an intake utilizing a thru flow screen is less than that of a dual flow. Based on the type and quantity of debris, the higher first cost may be offset by the reduced operating costs generally associated with dual flow designs.

CENTER FLOW SCREEN

The center flow screen is a variation of the dual flow traveling water screen that has a single entry/double-exit flow pattern. Water enters the center flow screen through an opening in the center of the screen frame and exits outward through both the ascending and descending baskets.

These screens are available for channel depths up to 34 ft (10.4 m), and most center flow screens have semicircular, or involute-shaped screen baskets. These shapes provide approximately 60% more screening area per basket and aid in the retention of debris during their ascent. Additional retaining plates may be added to the baskets to further enhance debris-carrying ability.

The radius of the chain guide track in the foot terminal may be more than twice that of the headsprocket, further increasing the submerged screen area.

The center flow screen has a debris spray located external to the screen, providing a downward debris discharge as the baskets reach their uppermost elevation.

THRU FLOW–TO–DUAL FLOW RETROFIT

Many existing thru flow screen installations can be modified or retrofit to use a dual flow traveling water screen. The modification consists of the installation of a special wall plate mounted perpendicular to the flow in

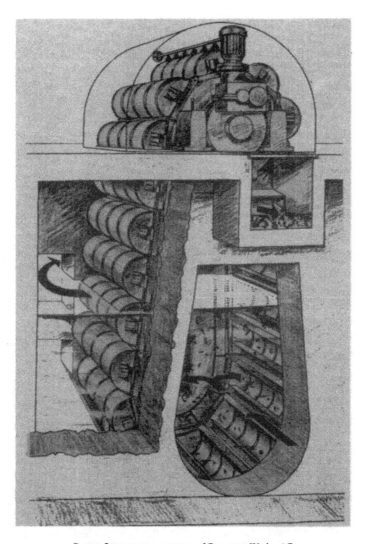

Center flow screen, courtesy of Passavant-Werke AG.

31

Center flow screen, courtesy of Bamag GmbH.

32

Center flow screen, courtesy of Brackett Green.

33

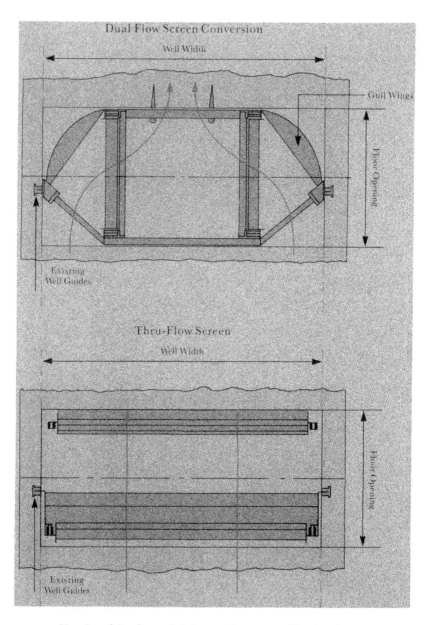

Plan view of thru flow to dual flow retrofit, courtesy of Brackett Green.

place of the existing screen. The dual flow screen is then lowered into the well, with baskets parallel to the flow, on the upstream side of the wall plate. An inlet opening in the wall plate allows screened water to pass to the pumps. An alternative arrangement uses a specially constructed screen mainframe that includes a wall plate made as an integral part of the screen frame with extensions or "wings" that fit into the existing embedded guides.

A thru flow to dual flow retrofit provides increased flexibility and may allow the operator to solve a specific operational problem. If an existing thru flow screen can be replaced with a dual flow unit of a similar basket width, the resulting increase in available screening area will provide a reduced velocity through the mesh. The user may also elect to install a screen with finer mesh, or a narrower basket width, without increasing the velocity.

SCREEN VARIATIONS

PLATFORM-MOUNTED

Double-entry dual flow screens can be mounted from a platform or pier that is supported by pilings, eliminating the need for a concrete channel or

Platform mounted screen arrangement, courtesy of FMC Corp.

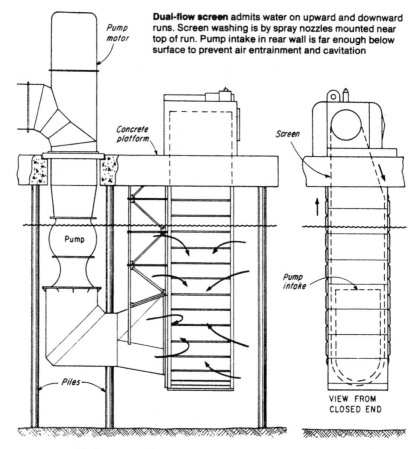

Dual-flow screen admits water on upward and downward runs. Screen washing is by spray nozzles mounted near top of run. Pump intake in rear wall is far enough below surface to prevent air entrainment and cavitation

Platform mounted power plant intake, courtesy of FMC Corp.

well. The flow pattern of this screen is similar to other double-entry screens: raw water enters the screen through the ascending and descending baskets and exits through an opening in the center of the screen structure. The screen exit opening is connected directly to the pump suction by duct-work.

This configuration may result in considerable intake construction cost reductions, and the openness of the design allows relatively easy escape for marine life.

INCLINED TRAVELING WATER SCREEN

A thru flow traveling water screen can be mounted at an inclination to enhance the screen's debris carrying capacity.

These screens are usually inclined ten to fifteen degrees from vertical, in the direction of the flow, although some screen designs may allow inclination of up forty-five degrees. This angle is often accompanied by an eight degree inclination of the wire mesh within the basket frame, resulting in a total mesh inclination of eighteen to twenty-three degrees. The placement of the screening surface at this angle increases the debris retention ability of the screen and helps prevent large solids from falling off the basket shelf as the baskets are lifted out of the flow.

The use of an inclined traveling water screen should be considered where very large debris loads are regularly expected or where the size and type of debris makes debris retention on vertical screens difficult. If regular debris loads in excess 15 ft³ (0.4 m³) per hour per foot of screen width are anticipated, an inclined screen may be required.

The inclination of the screen framework requires a rear debris discharge. The discharge point is located on the downstream side of the screen floor opening and is slightly cantilevered over the debris trough. This lessens the possibility of debris falling between the trough and the descending baskets, into the screened water.

Inclined screens may experience increased chain wear and somewhat higher operating and maintenance costs due to the chain rollers' operation in the inclined guide tracks.

Inclined screen installation, courtesy of Brackett Green.

BELT SCREENS

A belt screen is a simplified version of a thru flow traveling water screen. Belt screens can be used in light-duty screening applications, with channel widths to 6 ft (1.8 m) and channel depths to 20 ft (6 m).

Belt screens consist of an endless wire mesh belt in place of the individual screen baskets used in conventional traveling water screens. The belt may operate over a rubber lagged head pulley and slatted foot pulley, or it may be chain-mounted and operate over head- and footsprockets. Belt screens may be installed vertically or at inclinations of up to forty-five degrees.

Belt cleaning is accomplished by a high-pressure spray wash system, similar to that used on conventional traveling water screens. To increase the screen's lifting capacity and assist in debris removal, auxiliary lifting shelves can be attached to the belt at regular intervals.

Additional information on other belt screens can be found in Chapter 7 of this book.

Belt screen installation, courtesy of FPI Co.

Belt screen installation, courtesy of FPI Co.

Vertical belt screen.

SCREEN COMPONENTS

BASKETS (TRAYS)

Traveling water screen baskets, or trays, are the individual structural frames within which the screen mesh is mounted. The endurance and screening efficiency of a traveling water screen is largely dependent upon its basket design and construction. Baskets may range from 2 to 14 ft wide (0.6 to 4.3 m), with 10 ft (3 m) considered the industry standard for thru flow screens and approximately 6 ft (1.8 m) considered the standard for dual flow screens.

Baskets should be designed to maximize the available screening area and provide sufficient structural strength to withstand the loads imposed by a differential headloss of up to 5 ft (1.5 m) or more. Applications that routinely experience high debris loading should consider a design capable of withstanding the loads imposed by a 10 ft (3 m) headloss.

The basket's horizontal frame members act as beams to support the loads acting upon the basket and are designed to provide a seal with adjacent baskets to prevent debris from passing between them. This seal point is located on the centerline of the screen chain to insure that a constant clearance of 1/8″ to 3/16″ (3 to 5 mm) is maintained between adjacent baskets, even as they are rotated through the boot section.

Typical screen basket (tray) assembly, courtesy of FMC Corp.

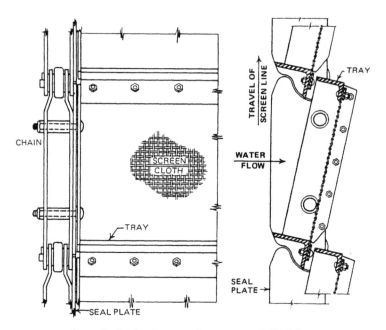

Screen basket (tray) construction, courtesy of FMC Corp.

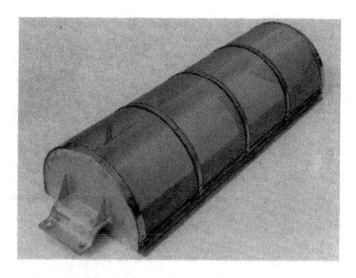

Center flow screen basket assembly, courtesy of Passavant-Werke AG.

42

The horizontal frame members, also called "lips" and "rails," are constructed of 1/4" or 3/8" (6 to 9 mm) thick, commercially rolled shapes (angles) or die-formed plates. These members form overlapping seals that prevent fish or debris from passing, even as the baskets flex around the sprockets. The lips are fixture-welded to curved end plates that seal the sides of the baskets as they revolve through the boot section.

Some manufacturers offer basket designs that utilize neoprene sealing strips between adjacent baskets to provide increased protection when small mesh openings are required. The lower, or trailing horizontal frame member also serves as a debris lifting shelf. This shelf assists in the retention of debris that does not cling to the screening media.

Wire mesh screen panels are assembled to the basket frame using clamp bars and carriage bolts. Vertical supports, spaced at 18" to 36" (457 to 914 mm) centers, are located on the downstream side of the mesh to assist in its support and minimize deflection. Mesh panels are often installed in the basket frame at a five to fifteen degree inclination to aid in debris retention.

Nonmetallic Baskets

Baskets fabricated entirely of nonmetallic materials have been used successfully since the mid-1980s. Fiberglass and fiberglass reinforced plastic components may be compression molded, pultruded, or extruded to form the required cross section.

The obvious advantages of nonmetallic baskets include greatly reduced weight (up to 50% less than steel) and increased corrosion resistance. The reduced weight results in significant reductions in wear on the chain, bearings, and other components. It also allows for easier installation and removal and facilitates operation at faster basket travel speeds.

Nonmetallic basket assembly, courtesy of Screening Systems International.

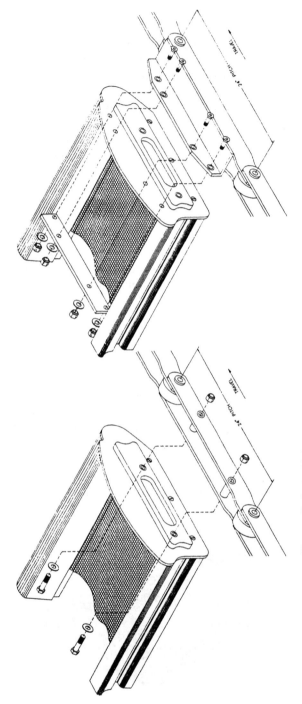

Nonmetallic basket/chain attachment methods, courtesy of Screening Systems International.

44

The strength, rigidity, and durability of the nonmetallic baskets currently available may vary dramatically among manufacturers. Manufacturers should be consulted regarding the specific performance characteristics of their nonmetallic baskets, especially their ability to withstand the design differential headloss conditions and the resulting flexural stresses encountered in screening operation.

Most existing traveling water screens utilizing steel or stainless steel baskets can be retrofit with nonmetallic basket frames. Some nonmetallic basket designs may provide less open area than their steel counterparts, which may adversely affect water velocities through the mesh.

Basket Attachment

There are several methods that may be used to attach baskets to the screen chain. Most methods are proprietary to individual manufacturers and include

- bolts passing through the chain sidebars and basket end plates
- bolting end plates to attachments welded on chain sidebars
- bolting basket "shelf" plates to a chain attachment pin

Auxiliary Backup Beam

If high differential headlosses are anticipated or if basket widths are greater than 10 ft (3 m), a center backup beam support may be considered. This system utilizes a vertical beam, with a stainless steel wear pad, located immediately behind the baskets and supported by the screen's main structural framework. This beam serves to support the basket frames as they deflect under the loads of a differential headloss. This system is most effective if the individual baskets are also equipped with a wear shoe. The wear shoe, manufactured of a low-friction material, will ride up the beam, minimizing basket deflection.

The use of a backup beam is intended to provide an added degree of safety when unusual operating conditions occur. Its use is not recommended for normal operation. The screen manufacturer should be consulted for basket widths greater than 10 ft (3 m).

Auxiliary Lifting Lips

Several types of auxiliary lifting lips are available to solve special debris-handling problems. Manufacturers offer auxiliary lip designs that can double or triple the debris-handling capacity of each basket. These lips may be toothed or plain, curved or flat, and may be bolted or welded

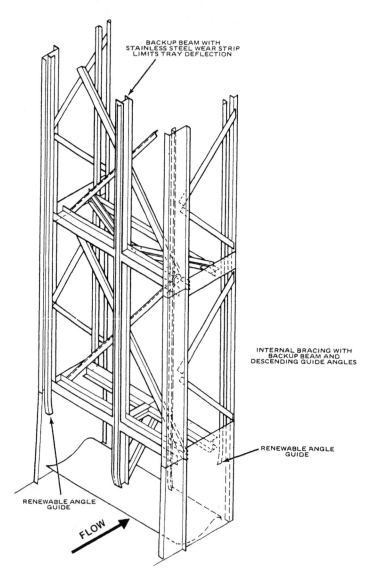

BACKUP BEAM WITH
STAINLESS STEEL WEAR STRIP
LIMITS TRAY DEFLECTION

INTERNAL BRACING WITH
BACKUP BEAM AND
DESCENDING GUIDE ANGLES

RENEWABLE ANGLE
GUIDE

RENEWABLE ANGLE
GUIDE

FLOW

Auxiliary backup beam arrangement, courtesy of FMC Corp.

to the existing lower debris shelf. For extreme debris loading requirements, a second auxiliary lifting lip can be mounted to the center of each basket.

MESH

Woven wire mesh is the most commonly used traveling water screening media. It is available in a variety of wire diameters and weaves, with square or rectangular openings of virtually any size, and can be manufactured to suit most material requirements. A typical traveling water screen uses a plain or double crimped weave, with a 3/8 " (9 mm) square opening and 0.080 " (2 mm) diameter type 304 stainless steel wire.

The typical range of mesh openings is 1/4 " to 3/4 " (6 to 19 mm), although some installations have successfully used mesh with openings smaller than 1/16 " (2 mm). If a fine [< 1/4 " (6 mm)] mesh is required, careful consideration should be given to the water velocity through the wire. Even if a fine mesh screen is designed within the normal, 1 to 2.5 fps (0.3 to 0.8 m/s), velocity range, it should be noted that the finer openings will retain more debris, more quickly, which may lead to a rapid buildup in headloss across the screen. A fine mesh may be impractical with many traveling water screens because their basket and framework designs are not capable of maintaining a seal less than 3/16 " (5 mm) between adjacent baskets or between the ends of the baskets, the boot section, or framework.

The individual strands of wire that make up the screen mesh generally have the smallest cross-sectional area of any part of the traveling water screen. It is important that they are as thin as practical, so that the maximum open area of the mesh can be realized. However, if the wire strands are too thin, they may be vulnerable to corrosion, damage, or puncture. The use of stainless steel or monel wire material is recommended over carbon steel, or galvanized carbon steel, in most applications.

An increasing number of installations are using a synthetic polyester monofilament screening fabric. This material is relatively rugged, corrosion resistant, inexpensive, and much lighter than stainless steel wire.

Although square woven mesh is most common, the use of a rectangular or oblong mesh opening may have advantages in some applications. A rectangular mesh opening of 1/8 " by 1/2 " (3 by 13 mm) has an open area greater than a mesh with a 3/4 " (6 mm) square opening of the same diameter wire. In addition, debris may be removed more efficiently from the oblong mesh because it tends to form a mat rather than stapling to the mesh.

Screens used in wastewater or CSO applications may utilize molded polyurethane mesh panels with round, tapered holes to reduce problems with debris stapling and aid in effective cleaning.

Interchangeable Mesh Panels

Baskets can be furnished with quick-change mesh inserts that allow the operator to replace the standard mesh with one better suited for seasonal operating requirements. For example, fine mesh screen inserts may be used to prevent the entrainment of fish eggs and larvae during the summer breeding season. After the breeding season, the original larger mesh screen inserts can be reinstalled.

SCREEN CHAIN

The chain that carries the basket assemblies is a heavy roller chain. Thru flow screens are usually provided with a 24" (610 mm) pitch chain, and dual flow screens usually use an 18" (457 mm) pitch chain.

Standard duty screen sidebars are 3/8" (9 mm) thick and 2-1/2" or 3" (64 or 76 mm) wide, fabricated from carbon or stainless steel. Heavy-duty, or very deep, screens may have 1/2" (13 mm) thick sidebars. European designs generally incorporate straight sidebar chain, while most United States manufacturers offer designs with double offset chain sidebars.

Chains are usually furnished with roller joints having heat-treated pins, rollers, and bushings that require periodic grease lubrication. Although some freshwater installations use carbon steel materials for chain joints, the use of a heat-treated series stainless is most common for both freshwater and seawater applications. Problem installations may consider using a precipitation hardened stainless steel or monel chain joints.

Nonlubricated chain designs are available from most manufacturers. These chains usually incorporate a cast chrome or nonmetallic (e.g., molded nylon roller) and hardened stainless steel pins and bushings.

Yet another chain design utilizes a labyrinth seal that helps retain lubricant inside the roller/bushing joint. This chain is reported to require lubrication after every 500 to 1000 hours of operation.

Some manufacturers offer chain links equipped with a glide block arrangement. A replaceable glide or wear shoe can be mounted to the outside sidebar of each chain link to minimize lateral movement and assure that the chain tracks properly.

FOOTSHAFT/FOOTSPROCKET

Thru flow traveling water screens are equipped with a pair of footsprockets, or footwheels, mounted on a shaft in the boot section of the screen frame. The maintenance and inspection of the footsprockets are difficult and costly, and their proper operation is critical to the performance and reliability of the screen.

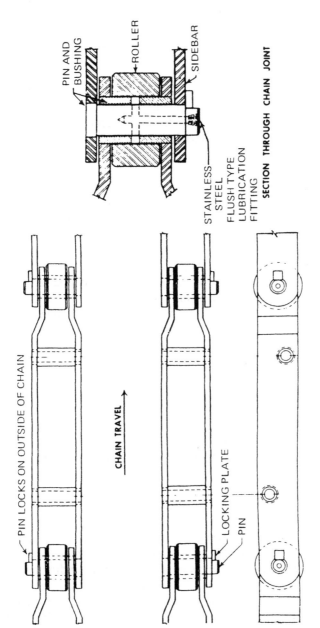

PIN AND BUSHING

ROLLER

SIDEBAR

STAINLESS STEEL FLUSH TYPE LUBRICATION FITTING

SECTION THROUGH CHAIN JOINT

PIN LOCKS ON OUTSIDE OF CHAIN

CHAIN TRAVEL

LOCKING PLATE

PIN

Screen chain, courtesy of FMC Corp.

49

Two shaft configurations are available: a "live" shaft that revolves within two bearing blocks and is fitted with keyed sprockets and a fixed "dead" shaft with bushed sprockets revolving on it. The bushing material for live shaft designs is either bronze- or a graphite-impregnated, molded canvas laminate, and the bearing block for standard dead shaft designs is usually manufactured of a self-lubricating bronze material.

Applications that require continuous operation or anticipate high levels of silt or grit should consider an alternate material selection for the foot-shaft/footsprocket bearing material. An alternative that has been used successfully in such abrasive applications is centrifugally cast chrome-nickel-boron alloy bushing and sleeve arrangement. This design utilizes a cast bushing operating over a similar sleeve that has been press fit onto the shaft. This material is considerably more expensive that the previously mentioned standard materials, but the expense can usually be justified.

A split shaft arrangement will reduce labor costs when the inevitable repair or replacement of footshaft components is required. The footshaft may be fabricated in two or three pieces, with each piece held together with ribbed compression couplings. If removal of the footshaft is required, it can first be disassembled within the screen and removed in sections. Footshafts may be fabricated of solid steel shafting or stainless steel tubing in diameters ranging from 2-7/16" to 3-7/16" (62 to 87 mm).

"Roll-Around" Foot Terminal

Most dual flow screens are equipped with a "roll-around" foot terminal in place of the conventional footshaft/footsprocket arrangement. The screen chain is guided in a 180 degree arc through the foot terminal, or boot section, by a formed steel, or cast iron guide rail. This design eliminates many of the problems associated with permanently submerged moving parts.

HEADSHAFT ASSEMBLY

Like other drive components, the headshaft must be adequately sized to withstand the NEMA-rated stall torque of the motor. The headshaft diameter is sized based on the combined torsion and bending moments and the allowable torsional deflection. Headshafts are available of high carbon solid steel shafting material in diameters ranging from 3-7/16" to 6-1/4" (87 to 160 mm). Some manufacturers offer torque tube headshaft arrangements to insure that overload conditions do not result in torsional deflection and bending that could twist or skew the baskets.

Headsprockets for 24" (610 mm) pitch chain have six teeth and a pitch diameter of 48" (1220 mm). Space limitations on some existing and re-

Footshaft assembly, courtesy of Screening Systems International.

Roll-around dual flow foot terminal, courtesy of FMC Corp.

51

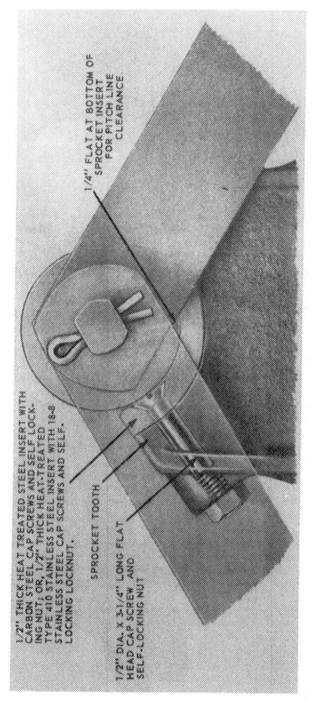

1/2" THICK HEAT TREATED STEEL INSERT WITH CARBON STEEL CAP SCREWS AND SELF LOCKING NUT, OR, 1/2" THICK HEAT-TREATED TYPE 410 STAINLESS STEEL INSERT WITH 18-8 STAINLESS STEEL CAP SCREWS AND SELF-LOCKING LOCKNUT.

1/4" FLAT AT BOTTOM OF SPROCKET INSERT FOR PITCH LINE CLEARANCE

SPROCKET TOOTH

1/2" DIA. X 3-1/4" LONG FLAT HEAD CAP SCREW AND SELF-LOCKING NUT

Headsprocket tooth insert, courtesy of Envirex, Inc.

52

Torque tube headshaft arrangement, courtesy of FMC Corp.

placement screens (and a very few new installations) may require a five-tooth sprocket design with a smaller pitch diameter. Headsprockets should be provided with replaceable tooth inserts of hardened stainless steel fitted to the sprocket teeth. Replaceable tooth pockets and segmented sprocket designs are also available to facilitate repair and removal of the headsprocket.

The centerline of the headshaft is typically located 3 to 4½ ft (1 to 1.4 m) above the operating floor level.

CHAIN TAKE-UP ASSEMBLY

The take-up assembly provides for adjustment of the chain tension by raising or lowering the headshaft. The headshaft assembly is supported on both ends by adjustable bearing blocks fixed to stainless steel or bronze take-up screws. The upper ends of the take-up screws are equipped with

Head shaft – Inside Drive with Torque Tube

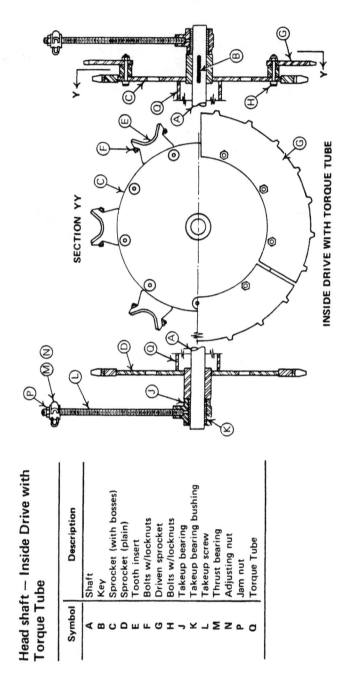

SECTION YY

INSIDE DRIVE WITH TORQUE TUBE

Torque tube headshaft arrangement, courtesy of FMC Corp.

Symbol	Description
A	Shaft
B	Key
C	Sprocket (with bosses)
D	Sprocket (plain)
E	Tooth insert
F	Bolts w/locknuts
G	Driven sprocket
H	Bolts w/locknuts
J	Takeup bearing
K	Takeup bearing bushing
L	Takeup screw
M	Thrust bearing
N	Adjusting nut
P	Jam nut
Q	Torque Tube

54

adjusting capstans and roller type thrust bearings and are supported by the screen headframe.

The standard traveling water screen take-up bearings are of the grease lubricated, bronze bushed type, although most manufacturers offer optional sealed, anti-friction roller bearings.

Inadequate chain tension is one of the primary causes of premature screen wear. Both over- and under-tensioning of screen chain are common operational problems. For this reason, most manufacturers offer equipment to assist or automate the chain tensioning procedure. Devices available include load cells with chain tension indicators, hydraulic tensioning systems, and spring-loaded adjustment equipment.

MAIN FRAMEWORK

The main framework of a traveling water screen is constructed of formed plate and structural shapes to provide freestanding support for the screen operating mechanism and chain guide tracks for the revolving baskets. Vertical flanges on each side of the frame mate with guideways embedded in or bolted to the channel walls to position the screen assembly in the channel and prevent the passage of debris around the unit.

Traveling water screens may be supported at the operating floor or channel bottom. If the screen is supported at the operating floor level, the headsection should be equipped with leveling screws. The leveling screws should be located over bearing plates that are fixed to the operating floor to allow for final screen adjustment after installation.

If the screen is supported at the bottom of the channel, the boot section merely rests on the bottom of the channel, with shims placed beneath it to level the screen shaft assemblies. The screen framework is made up of several frame sections that are bolted together with splice plates for easy handling and erection in the screen channel. These standard sections include the headsection or head terminal, upper intermediate section, intermediate section, and the boot section.

The headsection is that portion of the screen framework that is located above the operating floor level. The headsection supports the headshaft assembly and spray system and may be equipped with a machinery platform to support the drive unit. The headsection is usually fabricated of a minimum 1/4″ (6 mm) thick steel and should include provisions for attaching a hoist and lifting the entire screen assembly from the well. Headsections may also include optional jacking pads. These pads are welded perpendicular to headsection sideframe and provide a convenient location to insert a hydraulic jack to break the screen loose from the well and assist in its removal.

Access to the inside of the area of the headsection to inspect or maintain

Front/back view of two post screen frame, courtesy of Envirex, Inc.

56

Four-post screen frame assembled for shipment, courtesy of Screening Systems International.

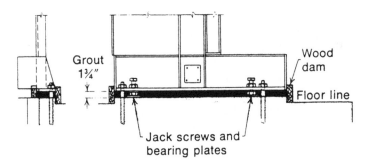

Headsection leveling arrangement, courtesy of FMC Corp.

Screen headsection, courtesy of Brackett Green.

the spray system and headsprocket tooth inserts is usually very limited. Most thru flow designs require removal of the splash housing and one or more baskets to access the spray system. Access for inspection and maintenance can be greatly facilitated by locating a manway in the side panel of the headsection framework and by installing a maintenance platform between the ascending and descending screen baskets.

The headsection is bolted to an upper intermediate frame section at approximately the operating floor level. This frame section joins the headsection to the remaining frame sections and usually varies from 5 to 14 ft (1.5 to 4.3 m) in height.

The intermediate framework is constructed in one or more standard sections measuring approximately 10 ft (3 m) high. These sections are usually fabricated of 3/8″ (9 mm) thick plate and are well stiffened and cross braced, forming a rigid box-like frame. The sideframes of the intermediate sections provide vertial guide tracks for the basket chain. Continuous vertical flanges on the outside of each sideframe are designed to fit into the guideways set in the concrete channel walls to prevent debris from passing around the screen frame. The intermediate sections should be constructed with provisions to allow the insertion of stop beams to support the frame at intermediate levels during screen installation, maintenance, or removal.

Frames of double-entry dual flow screens may be equipped with an automatic slide gate that can be activated during periods of extreme differential headloss. This gate is located on the screen side frame and, when activated, allows water to bypass the baskets and enter the center of the screen, relieving an extreme headloss condition.

Thru flow screens are available with a "two-" or "four-" post frame design. Two-post frames are designed to provide chain guide tracks for the upstream, ascending baskets only. To provide added rigidity for deep screen installations or other applications where it may be necessary to guide or confine the descending baskets, a four-post frame with downrun guide angles should be provided. Dual flow screens and screens with basket widths in excess of 10 ft (3 m) usually require four-post frame construction.

The boot section is the bottom-most frame section. The footshaft assembly is located in the boot section, with its centerline approximately 2-1/2 ft (762 mm) from the channel bottom. The boot section is equipped with a curved bootplate and track guide arrangement to seal the area under the screen as the descending baskets revolve around the footsprockets and begin their ascent. Most dual flow screens use a curved guide rail arrangement that eliminates the use of the footshaft, footsprocket, and bootplate.

For applications that experience high levels of sand or silt, the boot section may be equipped with a baffle plate that extends from the channel bottom to a point level with the footshaft. This baffle assists in keeping abrasive sand and silt in suspension and prevents sedimentation in the boot area.

The chain guide tracks are usually the only portion of the screen framework that are subject to wear under normal operating conditions. The life of the framework may be increased by utilizing bolted-type replaceable

Four-post screen construction, courtesy of Brackett Green.

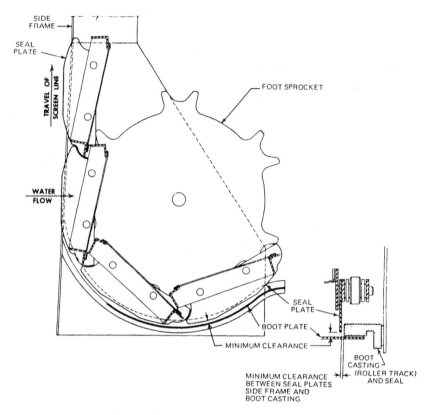

SIDE
FRAME

SEAL
PLATE

TRAVEL OF
SCREEN LINE

WATER
FLOW

FOOT SPROCKET

SEAL
PLATE

BOOT PLATE

MINIMUM CLEARANCE

BOOT
CASTING
(ROLLER TRACK)
AND SEAL

MINIMUM CLEARANCE
BETWEEN SEAL PLATES
SIDE FRAME AND
BOOT CASTING

Two-post boot section, courtesy of FMC Corp.

track angles or by lining the wear surface of the angles with replaceable steel or nylon wearing strips.

Redwood, cypress, nylon, or neoprene sealing strips may be used to further seal the area between the framework and the ends of the baskets. These sealing strips can be bolted to the upstream flanges of the mainframe to engage the basket endplates. Flexible sealing strips can also be installed to seal the boot section of a thru flow screen.

Frameless Designs

Some dual flow designs are provided without a conventional main frame or superstructure. These designs utilize vertical guide channels secured to the walls of the intake chamber that serves as chain tracks. This arrangement maximizes the use of the channel width but does not allow the screen to be removed as a unit for inspection or repair.

Four-post boot section, courtesy of Bamag GmbH.

SPRAY WASH SYSTEM

Debris is removed from the screen mesh by a system of high-pressure water sprays directed through the baskets into a debris trough. Most thru flow and dual flow traveling water screens use a front cleaning design that removes debris from the ascending baskets on the upstream side of the screen. Proper operation and maintenance of the spray system is essential to satisfactory screen performance.

Spray wash nozzle, courtesy of FMC Corp.

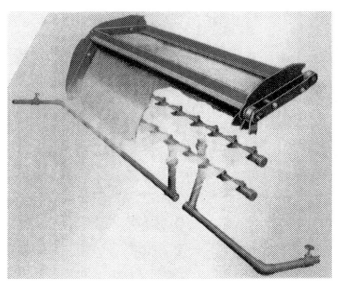

Spray system, courtesy of FMC Corp.

Spray water requirements vary based upon the screen width and the type and amount of debris anticipated. A typical spray wash system consists of a single row of venturi-type spray nozzles bolted or threaded to a spray pipe to provide a complete and slightly overlapping spray coverage along the width of the basket. If debris loading is very heavy or if debris is particularly tenacious, a second spray pipe and nozzle arrangement should be added.

Dual flow screen manufacturers usually recommend operating the screen spray system at approximately 35 psi (2.4 bar) and 18 gpm (1 L/s) per foot (0.3 m) of basket width. Thru flow screen manufacturers generally recommend spray system operation within a discharge range of 20 to 30 gpm (1.3 to 1.9 L/s) per foot (0.3 m) of basket width at a pressure range of 60 to 100 psi (4.1 to 6.9 bar), respectively. Field tests indicate that pressures of 60 psi (4.1 bar) are adequate for most thru flow applications.

The higher wash water pressure is recommended on thru flow screens because it is necessary to insure positive debris removal to prevent carryover problems. The design of a dual flow screen has an inherent protection against carryover, and the lower spray water pressure has proven satisfactory in most applications.

Excessive water pressure can also result in operational problems. Spray water pressures greater that 100 psi (6.9 bar) can result in a cascading effect when turbulence caused by the spray water hitting the inside of the splash housing results in debris being splashed or blown up onto clean baskets.

Most spray systems include an automatically operated valve that is actuated before the screen baskets are allowed to revolve. Some systems are also equipped with a pressure-sensing switch that prevents the screen from being rotated until adequate pressure is developed in the spray water line. The system should also be furnished with an automatic drain system to prevent damage from freezing when the screen is not in operation.

Spray wash water is usually obtained from the screened water on the downstream side of the traveling water screen. The use of an in-line, self-cleaning wash water strainer is recommended to minimize clogging of the spray nozzles. Strainers are also available with removable cartridges that can be removed for cleaning. Most strainers are equipped with filter openings that are one-third the size of the nozzle orifice.

A spray header and nozzle cleaning device may be provided with the screen spray system. One cleaning device consists of bristles mounted on a long rod located inside the spray header and connected to an external handwheel. The brushes can be manually rotated, by means of a handwheel, to periodically remove accumulated solids.

SPLASH HOUSING

Splash housings enclose the screen headsection and debris trough, contain the wash water spray, and direct it into the debris trough. Although semi-enclosed splash housings are available, totally enclosed housings are normally furnished for safety reasons.

Housings are usually constructed of molded fiberglass or light gauge stainless steel. The portion of the housing (usually the "front" housing) that directly covers the spray system and debris trough should be watertight, with appropriate seals and gaskets to prevent leakage. The front housing must have sufficient flexural strength to withstand the constant pressure of the spray water. Dependent on the materials of construction, the front housing should have a *minimum* thickness as follows:

- carbon steel—#10 gauge (0.1345") sheet
- stainless steel—#12 gauge (0.1046") sheet
- fiberglass—3/16" (5 mm) thick, molded

The rear housing is provided for safety purposes and to assist in the containment of wash water overspray. It should be easily removable or have a suitable opening to provide easy access to the baskets and chain for inspection and maintenance purposes.

All housing hardware and fasteners should be manufactured of stainless steel or other noncorrosive materials.

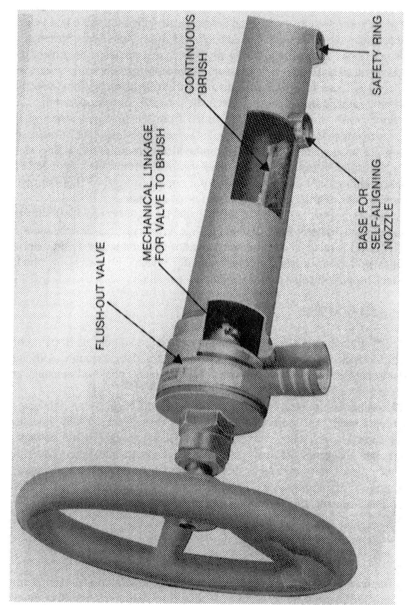

CONTINUOUS BRUSH

SAFETY RING

MECHANICAL LINKAGE FOR VALVE TO BRUSH

BASE FOR SELF-ALIGNING NOZZLE

FLUSH-OUT VALVE

Spray header with internal brush cleaner, courtesy of Spraco, Inc.

64

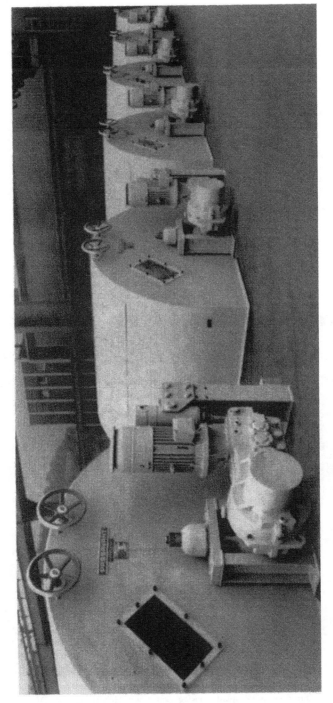

Center flow screen splash housings, courtesy of Passavant-Werke AG.

65

Thru flow screen splash housing, courtesy of Envirex, Inc.

Thru flow screen splash housing, courtesy of Hubert Stavoren BV.

DEBRIS TROUGH

The debris trough collects the spray wash water and refuse removed from the screen and transports it away from the intake area. Debris troughs should have a minimum depth of 6″ (152 mm), a minimum width of 1′-2″ (356 mm), and a minimum slope of 1/8″ in 12″ (3 mm in 305 mm) to assure proper flow in the trough. Multiscreen installations may use a common trough that spans the width of the intake. The portion of the trough between adjacent screens should be covered with checkered floor plate or grating.

Debris troughs are usually cast into the concrete intake structure during its construction, with the top of the trough located at the operating floor or deck elevation. Some installations utilize above ground troughs of steel, or fiberglass construction.

The edge of the debris trough is usually located 4″ to 10″ (102 to 254 mm) from the surface of the screening media. This gap should be bridged with a deflector plate to prevent water or debris from falling back into the channel between the debris trough and the baskets. The design and performance of this deflector are especially important on thru flow screens.

A typical deflector consists of a fixed, inclined plate fitted with an adjustable, flexible neoprene edge. The deflector can also be molded as part of the screen housing. Some manufacturers offer a deflector that is designed to follow the contour of the moving baskets to minimize the width of the gap.

Screens that utilize a front cleaned spray system and trough arrangement to clean ascending baskets may also utilize a deflector on the descending side of the machine. This secondary deflector can be installed without a spray system and with or without a trough. The deflector helps prevent debris not removed by the front spray system from falling into the downstream side of the screen. For example, horeshoe crabs often cling to the screen mesh and are not removed by the spray system. As they ride over the headsprockets and begin to descend, they may be discharged by gravity onto the secondary deflector.

DRIVE UNIT

Traveling water screen drive components include the electric motor, speed reducer, coupling, drive sprocket, driven sprocket, drive chain, and fluid coupling or shear pin overload protection system. Some manufacturers may provide a hollow shaft speed reducer mounted directly to the headshaft in place of the drive/driven sprockets and drive chain arrangement.

Drive unit components are selected to provide the capability of starting

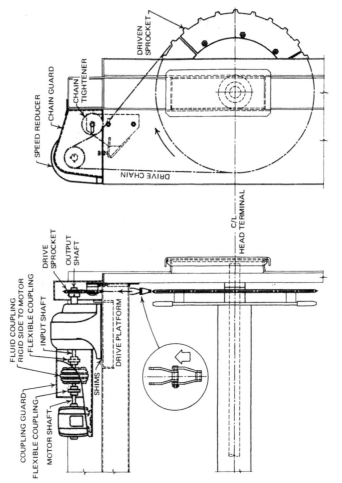

Typical drive arrangement, courtesy of FMC Corp.

DRIVEN SPROCKET

SPEED REDUCER

CHAIN GUARD

CHAIN TIGHTENER

DRIVE CHAIN

DRIVE SPROCKET

OUTPUT SHAFT

FLUID COUPLING RIGID SIDE TO MOTOR

FLEXIBLE COUPLING

INPUT SHAFT

COUPLING GUARD

FLEXIBLE COUPLING

MOTOR SHAFT

SHIMS

DRIVE PLATFORM

C/L HEAD TERMINAL

Shaft-mounted drive unit, courtesy of Brackett Green.

and/or operating the traveling water screen under a specified loading condition without exceeding the NEMA-rated starting and stall torque of the motor. The loaded conditions that are usually considered include starting the screen at a specified headloss and/or operating it continuously at a second specified headloss, at the maximum basket travel speed. Motor ratings for most applications range from 3/4 to 7-1/2 horsepower (0.6 to 5.6 kW).

The screen drive unit should be designed to start the screen under the loads imposed by a minimum 2.5 ft (0.8 m) differential headloss. Heavy-duty applications may require the screen to start under a 5 ft (1.5 m) headloss.

The industry standard basket travel speed is 10 fpm (3 ft/min). At this speed, clean baskets are introduced to the flow at a rate that can usually keep differential headlosses under control. If heavy debris loading conditions occur frequently, it may be desirable to operate at a higher basket speed. Typical high-speed operation is 20 fpm (6 m/min), although basket speeds as high as 45 fpm (14 m/min) have been provided.

Installations that are operated continuously or experience light debris loading may be operated at lower basket speeds. Reduced travel speeds of 2.5 to 5 fpm (0.7 to 1.5 m/min) will decrease wear and extend the life of screen components in direct proportion to the amount of speed reduction. However, these installations should consider the use of dual and multispeed drive units that provide the alternative of a faster basket speed for those occasions when debris loading may be higher than normal. Multiple basket speeds may be accomplished by using a two-speed or four-speed drive motor or a multiple speed transmission. Variable motor speed operation can be obtained through the use of a variable frequency, variable voltage, or motor speed control system.

SCREEN CONTROL SYSTEMS

The purpose of a screen control system is to hold the differential headloss and velocity across the screen at or near the optimum "clean screen" levels. Although some screens may be able to be operated manually, most installations require some type of automated controls to insure reliability and economy of operation. There are few limits to the versatility that can be designed into a screen control system.

Hand-Off-Auto

A simple screen control system may consist of a Hand-Off-Automatic (HOA) switch. In the "Hand," or manual position, the screen will continue to operate until turned to the "Off" position. When the screen is operated manually, the optional automatic control devices such as the time clock, limit switch, float switch, or differential level indicator will be overridden. In the "Automatic" position, the screen operation will be initiated by one of the optional control devices mentioned earlier.

Differential Level Controller

The purpose of a differential controller is to monitor the water level on the upstream and downstream sides of a screen and automatically perform a series of predetermined functions based on the differential water level. Differential controllers should be equipped with a minimum of two pressure switches or operational setpoints to indicate "low" and "high" differential levels.

Most differential controllers are designed to activate the screen drive motor at a preset level of approximately 6 " (152 mm). If the headloss is not relieved and continues to rise to a second preset level after the screen has been started, an alarm is sounded to warn of a potentially damaging headloss condition. Screens with two-speed drive motors may be switched to "high speed" at this higher differential level.

The controller should be designed to allow the screen to complete its operating cycle after the headloss drops below the "low" level. Traveling water screen controllers should be furnished with an adjustable timer to insure that the screen continues to operate for 1-1/3 revolutions after the headloss is relieved, to prevent excess debris from drying out on the baskets or rakes.

Controllers can be designed with additional level setpoints to shut down pumps at very high headloss levels or to initiate operation of other equipment or alarms. Control systems for traveling water screens should be designed to start the screen wash pumps and/or open screen wash water lines prior to starting the screen. It is also recommended that the system be equipped with a pressure sensing switch to insure that adequate line pressure is developed in the spray water system before the screen is rotated.

Pneumatic Differential Controller

A pneumatic differential level controller monitors headloss using two equally submerged vertical standpipes, or dip tubes, located on the upstream and downstream sides of the screen. A constant flow of air is delivered to the standpipes by means of a common pressure regulator to keep them purged of water. The difference in air pressure required to prevent intrusion of water causes a back pressure on both sides of a differential pressure switch. As long as the water level remains the same on both sides of the screen, the back pressure remains equal, and a differential level is not indicated.

As debris accumulates on the face of the screen, the downstream water level begins to fall. The change in water level is indicated as a difference in back pressure by the pressure switch. When the differential water level increases to a setpoint level, the pressure switch is closed and the screen activated.

To prevent the submerged ends of the standpipe from becoming clogged with marine growth or other debris, air pressure should be maintained. It is also suggested that the tubing be as straight as possible and installed vertically to allow cleaning with a push-rod.

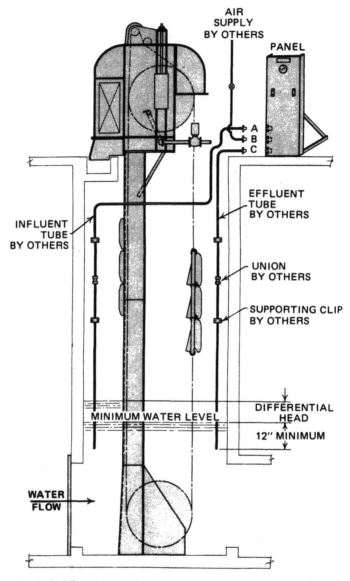

Pneumatic differential control system arrangement, courtesy of FMC Corp.

Ultrasonic Differential Controller

This system uses two ultrasonic transducers to measure the water level on the upstream and downstream sides of the screen. Both transducers transmit simultaneously, and a differential analog meter measures the time differential between the echoes to determine the water level differential. Some transducers require that they be mounted in a stilling well to minimize activation of the control unit by wave action.

Head Pressure Sensing System

This system utilizes pressure transducers that react to changes in water elevation by measuring the head pressure on the upstream and downstream side of the screen. The continuous level readings are sent to a differential control unit that controls the screen's operation as a function of the headloss differential.

TRAVELING WATER SCREEN DESIGN AND SELECTION

Traveling water screens must be designed to perform with a minimum of maintenance under a wide variety of operating conditions. They may be located on a river, reservoir, lake, or ocean and are subject to large fluctuations in flow conditions, debris loading, water depths, and salinity.

Determination of a traveling water screen's size is made by considering several factors unique to that application, including

- maximum, average flow
- maximum, minimum, average water levels
- number of screens
- wire mesh size
- velocity through mesh
- basket or channel width
- type of duty or service
- starting/operating headloss requirements

Selection procedures for thru flow and dual flow thru traveling water screens are similar. Unless specifically noted or sufficiently obvious, it may be assumed that the procedures describe herein apply to both types of screens. The selection of the screen model (thru or dual flow) is usually a matter of engineer or client preference, although an economic evaluation may help determine whether a thru flow or dual flow design is more cost-effective.

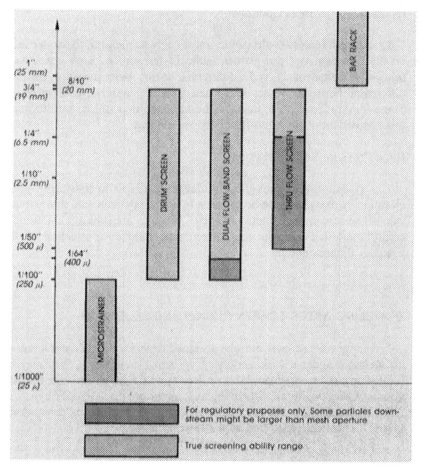

Screen selection chart, courtesy of Beaudrey Corp.

NUMBER OF SCREENS

The number of traveling water screens required for an installation is usually a matter of the engineer's preference. However, it is usually recommended that a minimum of two screens be installed. On the basis of operating and maintenance considerations, two narrow screens are recommended over a single wide screen. When two or more screens are used, the intake channels and screening equipment should be designed so that one screen can be taken out of service without adversely affecting the oepration of the remaining screen(s).

When possible, all screens at a single installation should be of the same model, manufacturer, and size. This will insure interchangeability and reduce the required spare parts inventory.

Future screening requirements should always be considered. If it is anticipated that additional screens may be required later, the size and location of the addition should be considered. It is often cost-effective to complete the civil work for future plant additions during initial plant construction.

INTAKE CHANNEL

Traveling water screens are installed side-by-side in individual concrete channels. Screens are usually preceded by a coarse bar rack to prevent large debris from damaging the screen. These bars are typically spaced to provide 2" to 5" (50 to 127 mm) openings.

When selecting a site for an intake, the engineer should first consider the

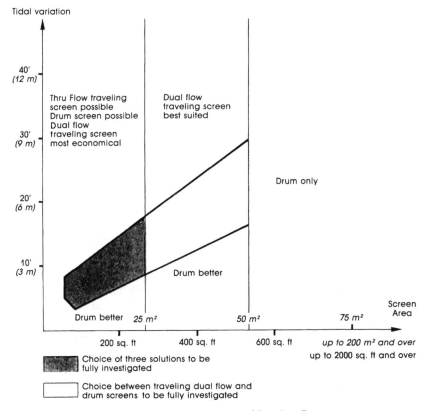

Screen selection chart, courtesy of Beaudrey Corp.

available maximum, minimum, and average (normal) water levels. In most locations, these water levels are fixed, and very little can be done about them.

It is recommended that the screen channel bottom be at or above the natural bottom of the water body. If a new channel bottom is dredged, problems due to sedimentation may arise as the artificial recess fills with silt or other sediment when the screen is not in use.

Although the minimum water level is normally the limiting factor in overall screen design (see "Velocities," p. 78), the maximum water level is usually considered when determining the screen's sprocket centers. It is necessary to locate the screen headsection and drive unit at an elevation that will insure that they will not be submerged during flood conditions.

CHANNEL WIDTH

Thru Flow Channel Width

Thru flow traveling water screens do not utilize the entire channel width as effective screen area. Ten to twenty percent of the channel area of a 10'-0" (3 m) wide screen may be restricted by the screen's framework and the guideways required to locate and fix the position of the screen in the channel. The interface of the guides and the screen frame also provides a labyrinth seal, preventing debris from bypassing the sides of the screen.

The use of vertical guides embedded or recessed in the channel walls is the preferred method of positioning thru flow screens. Embedded style guides consists of a "u-shaped" casting grouted into the channel walls so that the legs of the "u" are flush with the face of the wall. The screen is designed with a vertical guide flange as part of its main frame that will fit securely within the pocket of the embedded guide. A screen designed for use with an embedded style guide arrangement typically requires 1'-2" (356 mm) of the overall channel width. Thus, an 11'-2" (3.4 m) wide channel will be required for a traveling water screen with a basket width of 10'-0" (3 m).

Bolted style guides will most often be used where channels are existing or where their use is preferred for other design requirements. A bolted style guide consists of steel angles or a formed plate, bolted directly to the face of the channel walls. The protruding legs of the guide form a pocket to contain the frame guide flange. A bolted style guide arrangement will typically require 1'-8" (508 mm) of the overall channel width.

Dual Flow Channel Width

The effective screen width of a dual flow traveling water screen is parallel to the direction of the flow and has little bearing on channel width. The

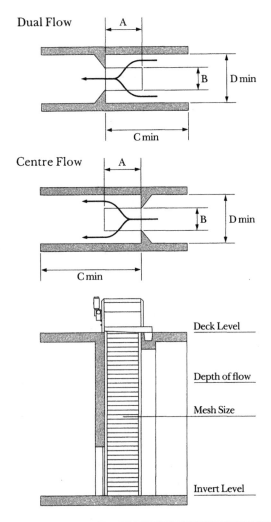

Depth of flow in metres	Capacity litres/sec					
1.0	220	440	-	490	-	-
1.5	320	650	970	720	1080	1570
2.0	430	850	1280	950	1430	2030
2.5	530	1060	1590	1190	1780	2500
3.0	640	1270	1900	1420	2130	2960
3.5	740	1480	2220	1650	2480	3430
4.0	850	1690	2530	1890	2830	3890
Dim A	0.9m	1.4m	1.9m	1.4m	1.9m	2.4m
Dim B	1.3m	1.3m	1.3m	1.4m	1.4m	1.9m
Dim C	1.5m	2.0m	2.5m	2.1m	2.6m	3.4m
Dim D	2.0m	2.0m	2.0m	2.3m	2.3m	3.2m

Dual and center flow sizing chart, courtesy of Brackett Green.

minimum channel width required for dual flow screens is usually two times the headsprocket pitch diameter. Because of the variety of headsprocket sizes, screen configurations, and hydraulic considerations, it is recommended that manufacturers be contacted for final recommendations.

WATER VELOCITY

As with the well depth, the width of a traveling water screen is usually determined by considering the extreme operating conditions. This condition usually occurs when the maximum flow occurs during a period of minimum water depth. When these conditions occur simultaneously, the water velocity through the screen increases, and the differential headloss across the screen may reach dangerous levels.

In most applications, a design velocity range from 1 to 2 fps (ft/sec) (0.3 to 0.6 m/s) will result in efficient and satisfactory operation. However, it should be noted that, with all other things equal, a screen with a design through-flow velocity of 1 fps (0.3 m/s) will be twice as wide as one with a design velocity of 2 fps (0.6 m/s).

Some applications may require velocities as low as 0.5 fps (0.15 m/s) or justify velocities as high as 3 fps (0.9 m/s). The selected design velocity should be based on the consideration of a number of site-specific factors:

(1) *Expected frequency and duration of extreme operating conditions:* If the extreme operating conditions occur infrequently or for very short periods of time, it may be acceptable to design at a higher velocity.

(2) *Type and quantity of debris to be removed:* If debris loads are normally very light and consist of easy-to-remove debris, it may be acceptable to design at a higher velocity.

(3) *Fish protection requirements:* Some applications require velocities as low as 0.5 fps (0.15 m/s) to minimize impingement of fish or other marine life.

(4) *Continuous operation:* Screens that operate continuously may be able to justify higher velocities than those that operate intermittently.

The water velocity through the mesh is calculated as a function of the minimum water depth, maximum flow, and screen efficiency. Screen size selection is normally determined by solving for basket width according to the following general formula:

$$BW = Q/(LW)(POA)(V)(K)$$

where

BW = basket width, ft

Q = flow, gpm
LW = water depth, ft
V = velocity through mesh, fps
K = 396 for thru flow screens
= 740 for dual flow screens
POA = % mesh open area (below)

Square Opening	Square Opening	W&M Gauge	Percent Open Area
3/16″	5 mm	#18	63.9
1/4″	6 mm	#16	63.8
1/4″	6 mm	#14	57.4
3/8″	9 mm	#14	67.9
3/8″	9 mm	#12	61.0
1/2″	13 mm	#14	74.3
1/2″	13 mm	#12	68.3

If a wire mesh with a rectangular opening or size is selected that is not listed above, the procedure for calculating the percent open area of the mesh is described in Appendix 3.

It should be noted that this procedure for determining water velocity will yield the theoretical average velocity through the mesh. It does not consider velocity concentrations that may occur due to channel hydraulics, pump locations, or unique basket configurations. It is to be used as a guide to assist in the determination of overall screen size. All velocity calculations should be verified by the screen manufacturer.

Consult the manufacturer for all velocities greater than 2.5 fps (0.8 m/s) and basket widths larger than 10′-0″ (3 m).

DIFFERENTIAL HEADLOSS

The differential headloss across a traveling water screen is the difference in water level between the upstream and downstream sides of the ascending baskets. The resulting headloss occurs as the flow of water in the channel is restricted by the wire mesh and screen structure.

The headloss is related to the water velocity through the mesh in general accordance with formula:

$$\Delta H = (V^2/2g)SF$$

where

ΔH = differential headloss, ft
V = velocity through mesh, fps

$g = 32.2$ ft/sec^s

$SF = $ safety factor

A traveling water screen with a design velocity in the recommended range of 1 to 2.5 fps (0.3 to 0.8 m/s) will have a clean screen headloss of less than 2″ (50 mm).

The velocity through the mesh will increase as the wire mesh becomes fouled with debris. An increase in velocity results in an exponential increase in the headloss across the screen. It is therefore important that the design velocity of a traveling water screen be considered at the extreme operating condition described above and that it is low enough to be able to effectively handle a sudden surge of debris. The presence of a headloss indicates that a severe or potentially severe condition may be developing. Any headloss in excess of 3″ to 6″ (76 to 152 mm) should result in operation of the screen.

The size and frequency of differential headloss occurrences have a direct bearing on screen operating costs. Traveling water screens operate under the same principal of operation that all other pieces of mechanical equipment do; the more frequent and severe the operation, the higher the operating cost.

As the headloss increases, the water pressure on the upstream baskets increases at a rate of 62.4 lb/ft^2 (305 kg/m^2) of headloss area. This means that a 3′-6″ (1.1 m) headloss occurring on a 10-ft (3-m) wide traveling water screen will result in a pressure of more than 1 ton (907 kg) being applied to an otherwise balanced load. The increased pressure created by a headloss forces the basket chain against the chain track guides and increases the chain pull required to move the baskets.

The increase in chain pull resulting from a differential headloss produces a proportional increase in the horsepower required to operate the traveling water screen. The additional pressure caused by a headloss also accelerates wear of the screen chain, chain tracks, bearings, sprockets, and other moving parts.

Potential headloss conditions should be considered when designing or specifying any traveling water screens. Most traveling water screen manufacturers design their "stand duty" screens so that they are able to start under the loads imposed by a 2′-6″ (0.8 m) headloss. "Heavy-duty" applications, including those located on large rivers or in tidal areas or those thay may encounter heavy or unusual debris loads, should be designed to be able to start the screen under the loads imposed by a 5′-0″ (1.5 m) headloss. For extremely heavy-duty applications, it may be recommended that the screen be able to be started and operated continuously while undergoing a 5′-0″ (1.5 m) headloss.

A standard 10′-0″ (3 m) wide traveling water screen should be structurally capable of withstanding the loads resulting from a minimum 5'-0" (1.5 m) headloss, and many applications today require an ability to withstand the loads imposed by a 10'-0" (3 m) headloss.

OUTDOOR OPERATION

Traveling water screens are routinely installed outdoors with a minimum of weather-related problems. However, some precautions should be taken when any machinery is located where it is subject to severe weather.

Common precautions, such as the use of proper lubricants during periods of temperature extremes, should be obvious. The mode of operation may also address potential weather-related problems. For example, continuous or frequent operation of traveling water screens may prevent the buildup of ice on mechanical components.

Spray Wash System

The primary cold weather concern with a traveling water screen is the screen spray wash system. If sustained periods of very cold weather may be experienced, care should be taken to insure that the spray system is protected from freezing. The system should be equipped with provisions to automatically drain the screen spray pipe upon screen shutdown. Heat-tracing of the spray water line is also recommended.

Frazil Ice

Intake screening equipment is susceptible to damage resulting from the formation and buildup of frazil ice. The formation of granular ice crystals in turbulent, supercooled water is referred to as "frazil ice." Supercooled water occurs when the water temperature begins to drop and passes through the 32°F (0°C) point. At a temperature of less than 32°F (0°C), sometimes even a fraction of a degree less, tiny particles of ice form quickly and uniformly throughout the water mass. Frazil ice is extremely adhesive and will stick to any solid object, such as a basket or screen frame, that is at or below the freezing point.

One method that has been used to solve the problems associated with the accumulation of frazil ice is to heat the trash rack bars. Although it is necessary to heat the bars 0.2°F (0.1°C) above ambient, they are usually heated at least 2°F (1.1°C). Heating may be accomplished using resistive heating elements imbedded in the bars or inductive heating using alternative currents to produce heat in strategically placed wiring or by circulating heated fluids through hollow cores in the bars.

Some electric power plants discharge small amounts of warm condenser exhaust at the intake to heat the water above the temperaure at which frazil ice formation will occur. However, environmental considerations may prevent this procedure from being used in all intake locations.

Other methods that have been used with varying degrees of success include the use of plastic coatings to inhibit ice adherence, the use of nonferrous materials (e.g., wood) with lower heat transfer properties, and the use of log booms or breakwaters that calm the water and allow the formation of an insulating cover of ice.

FISH SCREENS

Section 316(b) of the Federal Water Pollution Center Act of 1972 requires that "the location, design, construction, and capacity of cooling water intake structures reflect the best technology for minimizing adverse environmental impact" for all industrial categories. The environmental impact most often associated with cooling water intake structures is the effect of impingement and/or entrainment of marine life.

The entrapment of fish and other marine life on the screening media is referred to as "impingement." Impingement usually results when organisms cannot escape the area in front of the screen because of the velocity of the intake stream. Impingement is usually fatal because it leads to exhaustion and/or physical damage of the marine life.

Entrainment is the incorporation of small organisms, including the eggs and larvae of fish and shellfish, into the intake system. Entrained organisms pass through the intake screen and may be damaged as the result of mechanical, temperature, or pressure stresses associated with other downstream equipment or systems.

There are several design considerations and modifications of conventional traveling water screening equipment that can significantly reduce adverse environmental effects of a raw water intake system. These modifications are known collectively as "fish screens." Some fish screen features and technologies are proprietary, and most are applied on a site-specific basis.

Every water intake is unique. Impingement and entrainment problems may be seasonal or related to a tidal condition, intake location, water velocity, or a particular species of fish. It is generally accepted that the most desirable features to minimize the adverse environmental affects of an intake utilizing travel water screens are those that provide

- approach and through-flow velocities <1 fps (0.3 m/s)
- open or short intake channels with "escape routes"
- small mesh openings
- provisions to gently handle impinged fish
- continuous operation

Each of these features will add to the cost of the intake construction and/or equipment operation. It is important to determine the feature, or combinations of features, that results in the most cost-effective fish screening system for each particular application.

CONTINUOUS OPERATION

Most conventional screen installations are designed for intermittent operation; however, an essential feature of any fish protection system is its ability to operate continuously.

It has been demonstrated that continuous operation of traveling water screens alone, without any special fish handling provisions, will reduce impingement and entrainment of marine life. This can be attributed to the fact that continuous operation will prevent accumulations of debris that result in increased velocities and a subsequent differential headloss. As headlosses and velocities increase, it is far more likely that fish will not be able to escape the screen area and will become impinged and die.

If a screen has been designed with special fish handling provisions, it is important that the screen be operated continuously to remove impinged fish as quickly as possible. The shorter that the organisms are subjected to the stresses of impingement, the greater are their chances of survival. If a traveling water screen is to be operated continously, careful attention should be given to the selection of the mechanical components. The following items warrant special consideration:

(1) *Footshaft bearing or bushing material:* Standard bushings may be replaced with special wear-resistant materials or designs.

(2) *Footshaft assembly design:* A split-shaft design will allow easier replacement or repair of footshaft assembly components.

(3) *Headshaft take-up bearing:* Some manufacturers recommend the use of antifriction take-up bearings as replacements for standard, bronze bushed bearings.

(4) *Chain tensioning system:* A chain tension indicator, or an automatic chain tensioning device, will aid in maintaining proper chain tension.

(5) *Drive unit:* The drive components must have a service factor that reflects continuous operation.

(6) *Basket travel speeds:* Two-speed operation will reduce wear. A slow speed can be used for normal operation, and a higher speed can be used during periods of high debris or fish loading.

(7) *Differential controller:* This controller will only operate screens at "high speed" when debris loading indicates that it is necessary.

(8) *Chain material:* The material selection of the chain pins, rollers, and bushings should provide corrosion and wear resistance.

Continuous screen operation will also require that a firm inspection and maintenance schedule be prepared and followed. In addition to the above components, a routine inspection of the basket attachment bolts, headsprocket tooth inserts, nozzle orifices, and drive unit is required.

INTAKE LOCATION

The location of an intake is the first consideration necessary when designing a system to minimize fish impingement or entrainment. Most existing designs have been based on engineering and economic factors with little regard to the protection of fish and other marine life.

Biological field studies may provide important clues in determining the most practical method of minimizing the adverse effect of intake location or the aquatic life in the water source. The type and location of the intake should be considered relative to the water level and topography of the site. Other factors that must be reviewed include the location of plant discharge, navigational routes, recreational areas, ease of construction, and aesthetics. Improper location of an intake may actually attract marine life. Many "conventional" designs hopelessly trap fish in long dead end channels. A seemingly minor relocation of an intake to another place on the same plant site and water body may avoid areas of high fish concentrations.

FRONT DISCHARGE FISH SCREEN

There are two popular and reasonably economical "fish handling" alternatives to conventional traveling water screens. The front discharge fish screen is one of two alternatives that have been used successfully in the United States.

The front discharge fish screen is equipped with watertight fish pans mounted on the lower shelf of each basket. As the baskets revolve and are lifted out of the water, fish and other marine life are retained in 1-3/4″ (44 mm) deep fish pans. The pans have been designed so that a low-pressure spray system mounted outside the basket line displaces the water in the pan and flushes its contents into a fish sluice trough located on the front, upstream side of the screen.

After the pans have been emptied, a conventional high-pressure spray washes debris from the wire mesh into a separate debris trough, also located on the front, upstream side of the screen. Many existing traveling water screens can be retrofit to provide these fish removal features.

REAR DISCHARGE FISH SCREEN

The rear discharge fish screen is also equipped with watertight fish buckets mounted on the lower trash shelf of each basket. These buckets

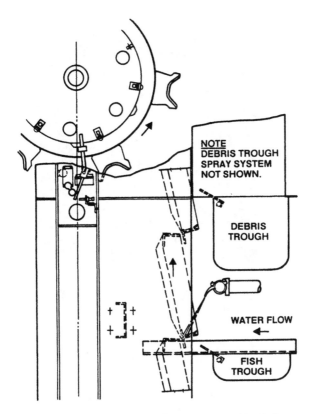

NOTE
DEBRIS TROUGH
SPRAY SYSTEM
NOT SHOWN.

DEBRIS
TROUGH

WATER FLOW

FISH
TROUGH

Front cleaned fish screen arrangement, courtesy of FMC Corp.

collect and retain marine life in 2″ to 4″ (50 to 102 mm) of water until the
baskets revolve over the headsprockets and spill their contents, with the aid
of gentle sluice sprays, into a fish trough for later release. The baskets continue past a high-pressure spray system to remove any remaining debris
into the debris trough for further disposal.

When considering a rear discharge fish screen, special attention should
be given to the basket and fish bucket design. It is extremely important that
the basket design provide a barrier-free discharge of marine life into the
fish trough. In most conventional basket designs, the screening media is
mounted at an eight-degree angle. This incline, coupled with some structural basket frame designs, may produce a ledge or pocket that may trap
or hinder the discharge of marine life as the baskets revolve around the
headsprockets.

Inclination of the entire screen frame will also facilitate rear discharge.
If any obstruction prevents a smooth rear discharge of marine life, the
baskets must be redesigned or modified with individual deflector plates.

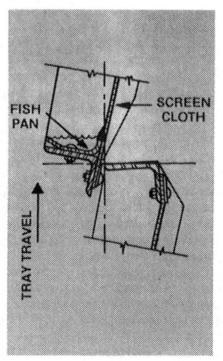

Front cleaned fish removal nozzle and bucket, courtesy of FMC Corp.

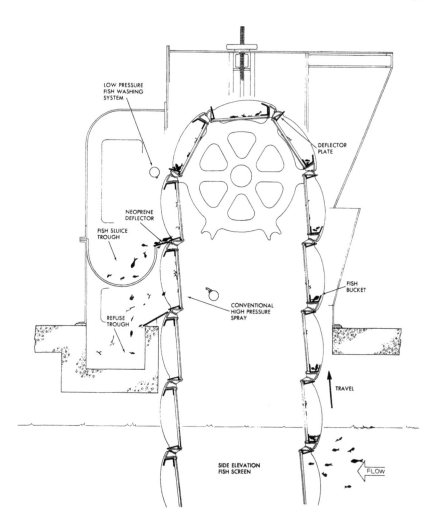

LOW PRESSURE
FISH WASHING
SYSTEM

DEFLECTOR
PLATE

NEOPRENE
DEFLECTOR

FISH SLUICE
TROUGH

CONVENTIONAL
HIGH PRESSURE
SPRAY

REFUSE
TROUGH

FISH
BUCKET

TRAVEL

SIDE ELEVATION
FISH SCREEN

FLOW

Rear discharge fish screen, courtesy of Envirex, Inc.

Some fish bucket designs may actually increase the damage to impinged fish. While submerged, water flowing over the leading edge of the fish bucket may cause a severe vortex that may repeatedly force fish into the wire mesh, causing them irreparable damage. Fish buckets should be designed to prevent formation of a vortex within them, to minimize stress on fish.

Many conventional traveling water screen designs can be modified to include a rear discharge fish recovery system. These modifications will include the addition of fish buckets on each basket, a low-pressure fish spray system, a fish sluice trough, and modifications to the housings. If the existing screen utilizes a front discharge spray system with a below grade debris trough, it may be necessary to increase the height of the screen headshaft to accommodate both troughs on the rear of the screen.

Modifications of existing screens are highly site-specific and should be carefully reviewed with potential manufacturers. In some cases, it may be more cost-effective to completely replace existing screens rather than to modify them.

FISH RETURN SYSTEMS

If a traveling water screen is fitted with a front or rear fish handling system, a fish return system must be provided to insure that any captured fish can be returned to the water body with a minimum of stress. The return system usually consists of a trough designed to maintain a water velocity of 3 to 5 fps (0.9 to 1.5 m/s) and a minimum water depth of 4" to 6" (102 to 152 mm). The trough should avoid sharp radius turns and should discharge slightly above the water level to avoid accumulations of marine growths on the submerged end of the trough. The trough should be covered with a removable cover to prevent access by birds or other predators.

One fish return system utilizes a special holding tank to separate the fish and debris removed by a screen before releasing fish. The tank is equipped with a herding device to isolate fish in one end of a tank, while debris is mechanically removed from the tank. After debris is removed, fish are released.

FLUSH MOUNTED FISH SCREEN

Many conventional thru flow and dual flow traveling water screens are located on the downstream end of an intake channel. Fish and other marine life entering the channel may become trapped in the channel as they are unable to escape the increased water velocity. They are impinged on the screen when they eventually tire and are no longer able to swim against the current to escape.

This problem may be overcome by mounting a thru flow screen so that the face of the screening surface is virtually flush with the shore line. Fish nearing the screen may be able to escape much easier by moving to the left or right of the screen face.

Coarse trash racks can be located out in front of and on the sides of the flush mounted screens to prevent damage from large debris, while preventing large enough openings for the fish to swim through freely.

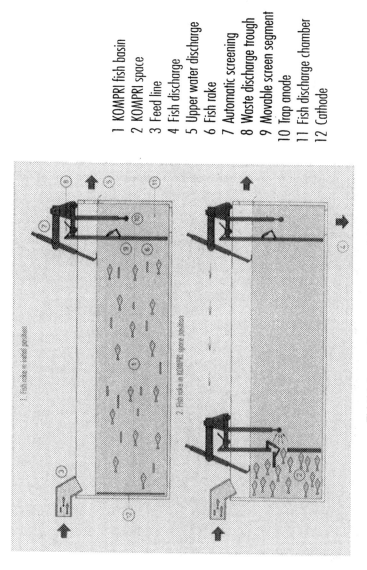

1 KOMPRI fish basin
2 KOMPRI space
3 Feed line
4 Fish discharge
5 Upper water discharge
6 Fish rake
7 Automatic screening
8 Waste discharge trough
9 Movable screen segment
10 Trap anode
11 Fish discharge chamber
12 Cathode

Fish/debris separation system, courtesy of Geiger GmbH & Co.

89

PLATFORM MOUNTED FISH SCREEN

Another design that eliminates the use of confining channels has a dual flow (double-entry/single-exit) screen mounted on a platform or pier. The screen simply "hangs" down into the water from the platform. Water passes through the screen baskets and into the center of the screen. It exists the screen through a port in the screen framework and into a plenum that may be connected directly to the pump suction. This screen may also be surrounded by a coarse bar grid to prevent damage from large debris.

ANGLED FISH DIVERSION SCREEN

Thru flow traveling water screens may be mounted at an angle to the flow to assist fish to an escape area by a velocity component of the flow.

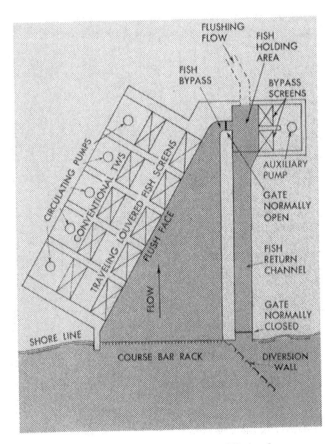

Angled fish diversion screen, courtesy of Envirex Inc.

Angled fish diversion screen, courtesy of Envirex Inc.

This arrangement may create an area of turbulence along the screen face that fish will avoid as they are directed to the escape bypass. As they reach the bypass, fish may be transported back to the water body by jet pumps, nonclog centrifugal pumps, airlift pumps, screw pumps, or a mechanical lift or "elevator" bucket.

An angled fish diversion system requires an extensive civil structure and may be cost prohibitive for most applications. It has also been demonstrated that this concept becomes less effective as the size of the intake increases.

BEHAVIORAL BARRIERS

There are a number of behavioral barriers that have been used to reduce impingement and entrainment of marine life at water intakes. These include electrical, sound, light, water jet, air bubble, stationary chain, and louver barriers. These methods have met with varying degrees of success, usually limited to specific species and sizes of fish and particular seasons of the year.

The use of one of these methods should be implemented only after thorough testing.

CHAPTER 4

Drum Screens

A drum screen is an in-channel screening device that consists of a series of wire mesh panels mounted on the periphery of a cylinder that slowly rotates on a horizontal axis. Drum screens are reliable screening devices utilizing very few moving parts.

Drum screens are widely used throughout Europe, Asia, and South America in electric power plants and other large water intakes. They may also be used to screen raw sewage at wastewater pumping stations, combined sewer overflows, and ocean outfalls.

The civil costs and the initial price of a drum screen installation may be higher than that of a traveling water screen installation. However, because of the design simplicity, the maintenance and operating costs of a drum screen installation are usually less than those of a traveling water screen.

Drum screens are usually furnished with woven wire mesh having square openings that range from 1/16″ to 3/8″ (2 to 9 mm), although some fine screening applications may utilize smaller openings. Drum screens used in wastewater applications may use plastic screening media panels with circular holes to facilitate cleaning.

The rolled steel rim of the drum is designed to mate with a correspondingly shaped angle bolted to the concrete intake walls to seal the drum. For finer screening applications, a formed neoprene seal may be fitted to the drum so that it contacts a low friction sealing face fixed to the chamber walls.

DOUBLE ENTRY DRUM SCREEN

A double entry drum screen is mounted in a concrete channel with its axis oriented perpendicular to the water's flow. Raw water enters the two open ends of the drum and flows through the wire mesh panels. Debris is retained on the inside surface of the wire mesh panels and elevated out of

93

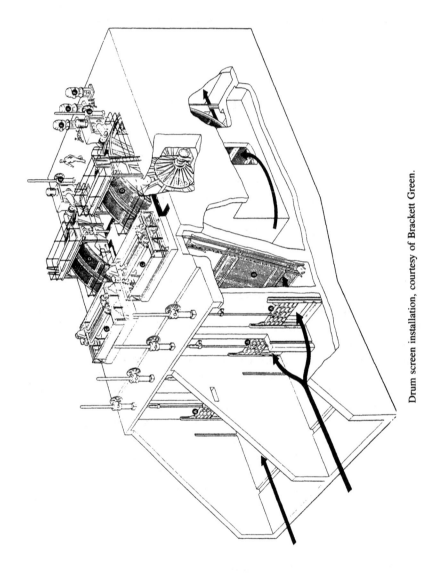

Drum screen installation, courtesy of Brackett Green.

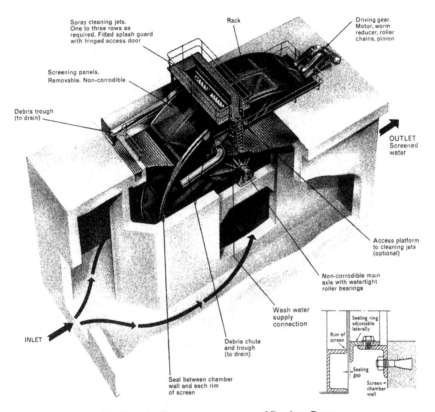

Spray cleaning jets.
One to three rows as
required. Fitted splash guard
with hinged access door

Rack

Driving gear.
Motor, worm
reducer, roller
chains, pinion

Screening panels.
Removable. Non-corrodible

Debris trough
(to drain)

OUTLET
Screened
water

Access platform
to cleaning jets
(optional)

Non-corrodible main
axle with watertight
roller bearings

INLET

Wash water
supply
connection

Debris chute
and trough
(to drain)

Seal between chamber
wall and each rim
of screen

Sealing ring
adjustable
laterally

Rim of
screen

Sealing
gap

Screen
chamber
wall

Double entry drum screen, courtesy of Brackett Green.

95

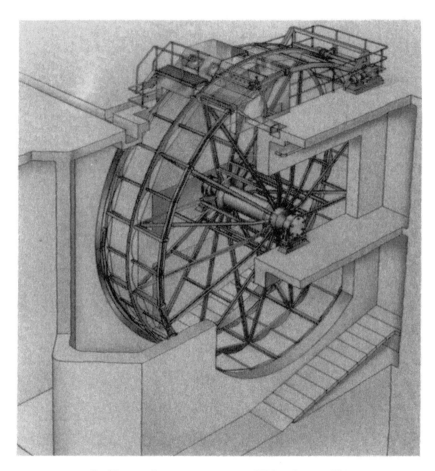

Double entry drum screen, courtesy of Hubert Stavoren BV.

Double entry drum screen drive gear, courtesy of Hubert Stavoren BV.

the flow as the drum rotates. Debris retention is aided by lifting shelves that are fitted to the inside of the drum at regular intervals. The debris is removed from the mesh into a debris trough at the top of the drum by a high-pressure spray wash system.

Double entry drum screens may measure up to 50 ft (15 m) in diameter, with effective screen widths of up to almost 14 ft (4.5 m).

The drum structure consists of a central steel shaft fitted with a cast iron hub to which a series of radial arms, or spokes, are attached. The central shaft rotates within sealed roller bearings mounted on joists that span the entrances to the drum. Structural steel cross members are fitted to the spokes to support the circular rims of the drum assembly. The circular rims form the framework to which the screening media is attached.

The drum is driven by means of a segmented gear rack that is centrally mounted to the periphery of the drum. A driving pinion mounted near the operating floor elevation engages the gear rack and rotates the drum.

SINGLE ENTRY DRUM SCREEN

Single entry drum screens, also called cup screens, are similar to the larger double entry drum screens. The major difference lies in their orientation to the flow and the fact that the downstream end of the drum is closed with a full-diameter backplate. The central shaft of the cup screen is located parallel to the flow and the open end of the drum faces the flow.

Flow enters the drum through its open end and exits outward through the wire mesh mounted on the screen's periphery. Debris is retained on the inside surface of the wire mesh panels and is elevated out of the flow as the drum rotates. A high-pressure spray wash system is used to remove debris.

The gear rack used to drive the screen is mounted to the backplate, and an adjustable thrust bearing is usually located on the downstream side of the screen's central shaft to absorb thrust loads.

Cup screens are available in 5 to 25 ft (1.5 to 7.6 m) diameters and widths of 1 to 8 ft (0.3 to 2.4 m). For flows of 20,000 gpm (1260 L/s) or less, tank mounted, package units are available.

PUMP SUCTION SCREEN

Another cylindrical screen design utilizes a slowly rotating wire mesh screen with a water backwash spray to prevent accumulations of debris

Single entry drum screen installation, courtesy of Brackett Green.

Single entry drum screen, courtesy of FMC Corp.

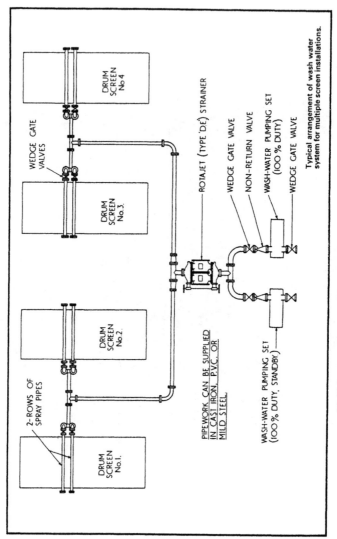

Multiple screen spray wash system, courtesy of Brackett Green.

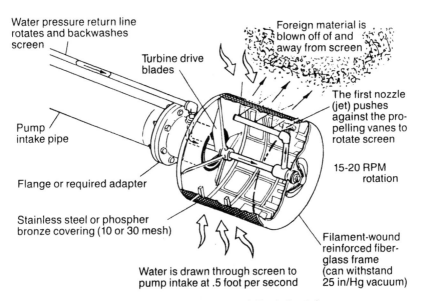

Water pressure return line rotates and backwashes screen

Turbine drive blades

Foreign material is blown off of and away from screen

Pump intake pipe

The first nozzle (jet) pushes against the propelling vanes to rotate screen

Flange or required adapter

15-20 RPM rotation

Stainless steel or phospher bronze covering (10 or 30 mesh)

Filament-wound reinforced fiberglass frame (can withstand 25 in/Hg vacuum)

Water is drawn through screen to pump intake at .5 foot per second

Pump suction drum screen, courtesy of Claude Laval Corp.

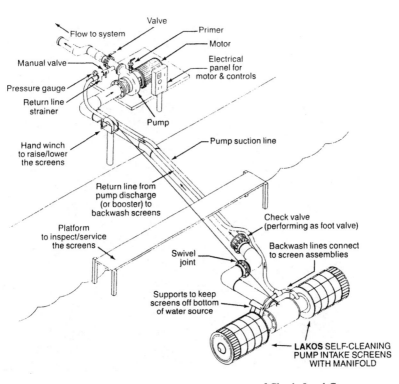

Valve

Flow to system

Primer

Motor

Manual valve

Electrical panel for motor & controls

Pressure gauge

Return line strainer

Pump

Hand winch to raise/lower the screens

Pump suction line

Return line from pump discharge (or booster) to backwash screens

Platform to inspect/service the screens

Check valve (performing as foot valve)

Swivel joint

Backwash lines connect to screen assemblies

Supports to keep screens off bottom of water source

LAKOS SELF-CLEANING PUMP INTAKE SCREENS WITH MANIFOLD

Pump suction drum screen arrangement, courtesy of Claude Laval Corp.

from a pump intake. These units are available with 10 and 30 mesh openings. Individual screen sizes range from 15″ to 24″ (381 to 610 mm) diameter and 9″ to 34″ (229 to 864 mm) long.

Twelve to 20 rpm screen rotation is accomplished as the 55 to 90 psi (3.8 to 6.2 bar) backwash water simultaneously removes debris and pushes against propelling vanes mounted inside the screen cylinder.

These units are not recommended for use in confined areas without any means of moving debris away from the screen area, or in currents over 4 fps (1.2 m/s).

DISC SCREENS

The revolving disc screen is a simple and compact screening device that is closely related to the drum screen. It consists of a flat disc covered with screening media that rotates about a horizontal axis, perpendicular to the water flow. As water flows through the submerged portion of the disc,

Disc screen, courtesy of FMC Corp.

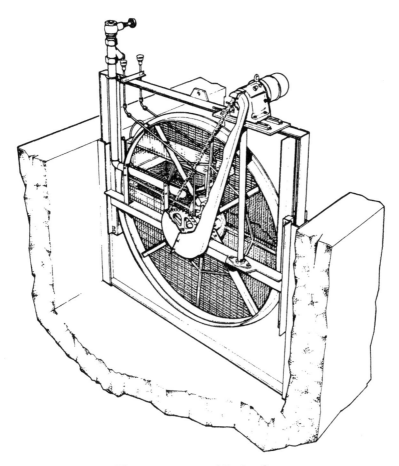

Disc screen, courtesy of Brackett Green.

solids are retained on the screening media. The rotation of the disc lifts the solids above the water surface where they are removed by a series of spray nozzles mounted on a spray header.

Disc screens are available in sizes ranging from 6 to 16 ft (1.8 to 4.9 m) in diameter with mesh openings as fine as 0.040 " (1 mm). Applications are limited to those with small fluctuations in water levels.

CHAPTER 5

Passive Intake Screens

PASSIVE intake screens are stationary screening cylinders positioned in a water body so as to take advantage of ambient currents and controlled through-screen velocities to minimize debris buildup. Most passive screen intakes have no moving parts and require no debris handling or debris disposal equipment.

Passive screens are usually mounted on a horizontal axis and oriented parallel to the ambient currents that provide necessary cleaning action. Most installations are designed for a maximum intake velocity of 0.5 fps (0.15 m/s) through the screen opening to minimize debris impingement on the screen surface.

The periphery of the passive screen cylinder provides a screening surface consisting of a trapezoidal "wedge wire" bar formed to maintain a uniform screen opening that typically ranges from 0.040" to 0.50" (1 to 13 mm). Screen cylinders may range in size from 12" to 84" (0.3 to 2.1 m) in diameter, and 12" to 96" (0.3 to 2.4 m) long and are usually designed to withstand a differential pressure equal to 10 ft (3 m) of water across the screen surface.

Most passive screens are fabricated of stainless steel or other corrosion-resistant materials. When biofouling may be a problem, screens can be fabricated of a copper-nickel alloy to resist fouling through a slow, steady release of copper that is toxic to potential encrustants.

Screens should be designed to provide uniform velocities across the entire screen surface. The elimination of areas of high localized velocities will prevent debris accumulations. Uniform flow also increases the hydraulic capacity and pump efficiency by reducing suction headloss.

The stress experienced by marine life during the mechanical handling involved with other screening processes is eliminated, and the finer openings and low velocities associated with passive screens result in their reduced impingement and entrainment.

Passive intake screen, courtesy of Wheelabrator Johnson Screens.

These are a number of intake configurations available to suit a variety of application requirements. Small installations may use a single screen, a single tee-unit with two screens, or a multi-tee system. Large installations may utilize multiple screen arrays manifolded to provide equal flow through each screen.

Multiple screens installation, courtesy of Wheelabrator Johnson Screens.

A tower arrangement may be provided with screens mounted at various elevations. This will allow the operator to select the level at which water will be drawn to accommodate seasonal changes in water level, quality, or temperature.

Passive screens can be connected directly to a pump suction, or they can be arranged for gravity flow to feed an on-shore wet well. The wet well may be a reinforced concrete caisson that is constructed on-site using the open-end caisson method of sinking to eliminate dewatering and extensive excavation, minimizing local environmental impact.

Periodic physical cleaning may be required due to debris accumulations or biofouling, especially on marine intakes. If frequent cleaning is antici-

Passive screen tower arrangement, courtesy of Wheelabrator Johnson Screens.

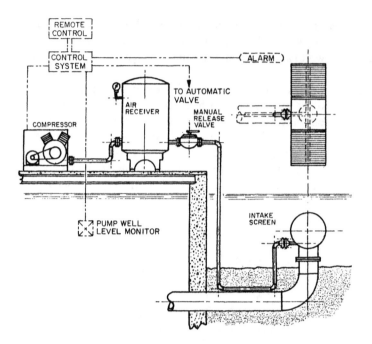

Passive screen backwash system, courtesy of Wheelabrator Johnson Screens.

pated, bulkhead mounting of the screens is recommended to provide easier access. Bulkhead mounted units are located over inlet ports on vertical rails to facilitate their removal. In locations with high-debris concentrations or where access to the screen is difficult, an auxiliary air burst or water backwash cleaning system may be furnished.

An air burst cleaning system consists of a factory-assembled package that includes an air compressor and accumulator, distribution manifold, control system, and individual screen air distributor. When cleaning is required, two to three screen volumes of air stored at 100 to 150 psig (6.9 to 10.3 bar) is relased rapidly, scouring accumulated debris from the screen and carrying it up and away from the screen.

RADIAL WELL INTAKE SCREEN

A radial well intake is an infiltration system that utilizes slotted pipes or screens placed horizontally in a naturally occurring sand or gravel aquifer. Most installations utilize several pipe screens that radiate outward from a vertical wet well or caisson. Water enters the system through the screen pipes and empties into the central caisson where it is removed by pumping.

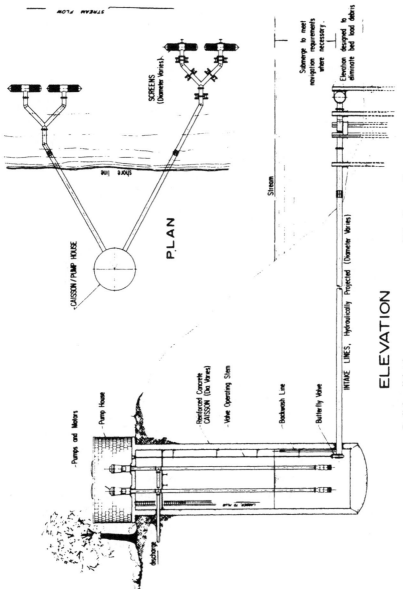

STREAM FLOW

SCREENS
(Diameter Varies)

shore line

CAISSON / PUMP HOUSE

PLAN

Stream

Submerge to meet
navigation requirements
where necessary .

Elevation designed to
eliminate bed load debris

INTAKE LINES, Hydraulically Projected (Diameter Varies)

Pumps and Motors

Pump House

Reinforced Concrete
CAISSON (Dia Varies)

Valve Operating Stem

Backwash Line

Butterfly Valve

Discharge

ELEVATION

Radial well intake arrangement, courtesy of The Ranney Co.

109

Radial well screen installation, courtesy of The Ranney Co.

The intake is constructed by installing a vertical caisson below the water table near a river or creek with radial wells jacked out into the surrounding aquifer. They are "developed" in the same manner as conventional wells.

The use of this intake system is limited to locations that have a permeable substrate and reasonably clean water. The maximum capacity of a single well is approximately 20,000 gpm (1260 L/s), although several wells may be placed in a group to provide increased flow.

Where a combination of appropriate geology and flow capacities exist, this intake competes favorably with conventional intake designs. This design also provides reduced operating expenses than conventional wells due to lower pumping heads and greater pumping efficiencies.

Radial well intake systems have the environmental advantage of segregating the marine life populations from direct contact with the screen mechanism or screen structure.

CHAPTER 6

Bar Screens

A bar screen consists of a stationary bar rack that is automatically cleaned by one or more power operated rakes. As a rake is operated up the face of the bar rack, it removes accumulated debris and elevates it out of the flow. At the top of the rake's operating cycle, debris is swept from the rake into a debris receptacle by a wiper mechanism.

Bar screens are most frequently used at the headworks of wastewater treatment plants to remove objectionable material that may interfere with, or not respond to, accepted methods of treatment. Because they are the first unit process in a treatment plant, they play an important role in determining overall plant performance. Bar screens also remove large solid objects, rags, and other debris that may clog piping and damage pumps or other downstream equipment at pump stations, flood control projects, and irrigation facilities.

Although some small plants still rely on manually cleaned bar racks, most modern plants employ mechanically cleaned bar screens to increase plant efficiency by removing debris accumulations as they occur. This reduces the possibility of sewage backups and grit sedimentation that would otherwise occur if a screen became plugged. It also eliminates the subsequent influent surge that would accompany the cleaning of a plugged screen.

Industrial plants often employ bar screens as a method of pretreating their wastewater before discharging plant effluent into a municipal sewer. In some industrial installations, the screens may be used to recover sufficient materials to pay for their installation and operation.

Bar screens are usually categorized according to the type of raking mechanism used to remove debris from the bar rack. There are bar screen design variations available to suit most screening requirements and operator preferences for channel widths measuring up to 14 ft (4.3 m) wide and depths of up to 80 ft (24 m).

The quantity of screenings removed by a bar screen varies as a function of the width of the openings in the bar rack. A typical municipal wastewater treatment application, with bar spacings that range from 1/2″ to 1-1/2″ (13 to 38 mm), will yield 0.5 to 10 ft³ of screenings per million gallons of wastewater.

RECIPROCATING RAKE BAR SCREENS

Reciprocating rake bar screens are automatic screening devices that use a single, reciprocating rake to clean a stationary bar rack in an action that simulates manual raking. The mechanism used to raise and lower the cleaning rake varies considerably among reciprocating rake bar screen designs.

The great increase in the popularity of these screens is due to the fact

Reciprocating rake bar screen, courtesy of Infilco Degremont.

that they have relatively few mechanical parts, and most designs have no moving parts permanently located below the designated maximum water level. This facilitates inspection and maintenance and means that most of the work required can be done at the operating floor level. Because so few parts are exposed to water, it may be economically feasible to specify all "wetted" parts be constructed of stainless steel or other suitable corrosion-resistant material.

The operating cycle of a reciprocating rake screen begins with the rake in the standby, or parked, position at the discharge elevation. The cleaning rake is lowered to the channel bottom with the rake teeth extended in a dis-engaged position. As the rake reaches the channel bottom, the teeth are pivoted in the direction of the flow until the teeth penetrate the upstream face of the bar rack. The rake is then lifted up the face of the bar rack to the discharge elevation where the screenings are removed by a pivoting wiper mechanism.

The typical rake travel speed for reciprocating rake bar screens is 20 ft (6 m) per minute. Some designs are able to use a two-speed drive motor that can provide a higher speed for the descending stroke of the rake. This arrangement will produce increased debris handling capability by adding a significant number of rake cleaning cycles per hour. The drive motor should be equipped with an integral motor brake to allow the rake carriage to be stopped at any operating level.

Reciprocating rake bar screens are generally used in applications with channel widths of 2 to 12 ft (0.6 to 3.7 m), although some units have been manufactured with effective rake widths of almost 30 ft (9.1 m). Maximum channel depths vary among screen designs, although a maximum depth of 25 ft (7.6 m) is recommended for most reciprocating rake applications. This is due to the time required to complete one cleaning cycle. Long cycle times may be prohibitive on deep installations or applications that experience high solids loadings.

Like the drive components, most of the screen framework is located above the water level, and the frame is usually anchored to the operating deck. The bar racks for reciprocating rake bar screens may be mounted at a zero to twenty-five degree inclination. Most reciprocating rake screens are of the front cleaned design, eliminating potential debris carryover problems, although several manufacturers offer back cleaned screen designs.

Some reciprocating rake bar screens are designed with the drive motor mounted on a carriage that reciprocates with the cleaning rake. When these screens are applied in installations that may experience extremely high water depths, the drive unit may be subjected to periodic submergence. Proprietary, submersible drive enclosures, or hydraulic drive units are available for these applications.

Discharge heights or headroom restrictions at some installations may require a screen to discharge at an elevation below grade. For these installations, a lift mechanism or conveyor may collect debris at an intermediate level and elevate it to the final discharge elevation.

COGWHEEL AND PIN RACK RECIPROCATING RAKE

The most popular reciprocating rake design uses a cogwheel and pin rack drive arrangement to operate the cleaning rake. The drive mechanism and cleaning rake assembly are mounted to a traveling carriage equipped with a pair of cogwheels. As the gear motor turns the cogwheels, the carriage travels up and down a fixed pin rack, raising and lowering the cleaning rake. Some manufacturers use a toothed track rail or gear rack instead of a pin rack.

The engagement and disengagement of the rake teeth with the screen bars is controlled by the position of the cogwheels on the pin rack. As the

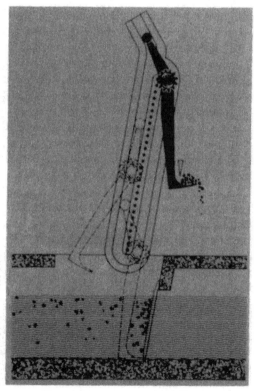

Pin rack bar screen operational diagram.

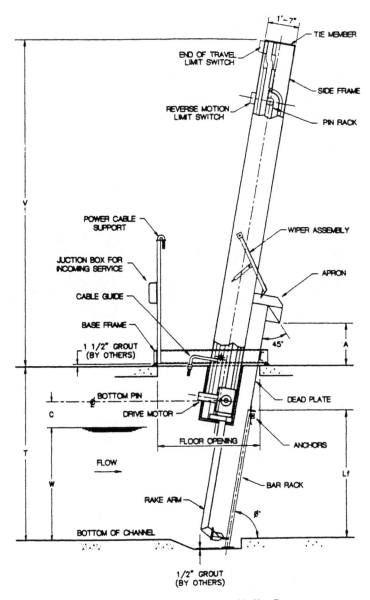

1'-7"

TIE MEMBER

END OF TRAVEL
LIMIT SWITCH

SIDE FRAME

REVERSE MOTION
LIMIT SWITCH

PIN RACK

POWER CABLE
SUPPORT

WIPER ASSEMBLY

JUCTION BOX FOR
INCOMING SERVICE

APRON

CABLE GUIDE

BASE FRAME

1 1/2" GROUT
(BY OTHERS)

45°

A

BOTTOM PIN

DEAD PLATE

C

DRIVE MOTOR

FLOOR OPENING

ANCHORS

T

FLOW

Lf

W

BAR RACK

RAKE ARM

Ø°

BOTTOM OF CHANNEL

1/2" GROUT
(BY OTHERS)

Pin rack bar screen arrangement, courtesy of Infilco Degremont.

Pin rack reciprocating rake bar screen installation, courtesy of FMC Corp.

cogwheels travel around the bottom of the rack, the cleaning rake rotates into the engaged position for the ascending, cleaning stroke. After the screenings have been discharged and the cogwheels travel around the top of the rack, the rake pivots to the disengaged position for the descending, return stroke. The rake arm is equipped with guide, or follower, rollers that travel within a guide track to fix the upper position of the rake arm relative to the screen bars.

If the rake encounters an obstruction during the cleaning stroke, the cogwheels stop turning, and the upward travel of the rake carriage is halted. The motor continues to operate, resulting in a tilting of the drive unit as it rotates slightly in a counterclockwise direction on the shaft. The tilting action compresses a coil spring, and the rotating movement causes the rake to disengage. The carriage can again begin its upward movement, riding over the obstruction. If the object still interferes with travel and the teeth are in the disengaged position, the coil spring is further compressed, activating a limit switch that stops the drive motor.

The follower roller guide tracks should be designed and manufactured to insure a smooth and continuous rake motion, to prevent screenings from

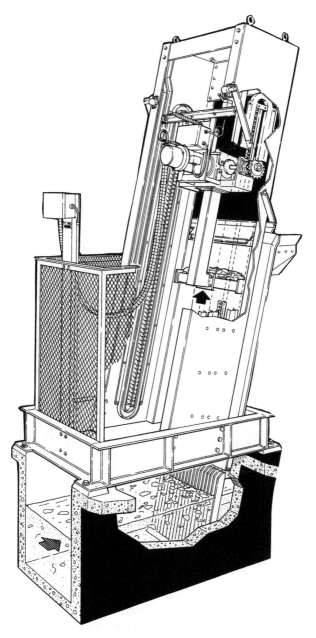

Pin rack bar screen arrangement, courtesy of Schloss Engineered Equipment.

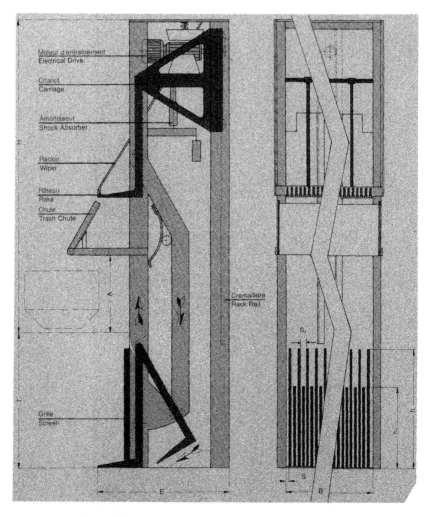

Back cleaned bar screen arrangement, courtesy of John Meunier, Inc.

being shaken off the rake, and to minimize wear and tear of the screen. This is especially important as the rollers pass through the track cam offset.

Headroom requirements for this screen design should be carefully considered. An estimate of the required headroom can be made by adding the vertical discharge height above the operating floor to the vertical depth of the bar rack. This sum will *approximately* equal the headroom required above the discharge elevation. The screen manufacturer should be consulted for actual headroom requirements.

These units are available with bar spacings to 1/4 ″ (6 mm) and options that include hydraulic drive units utilizing nonpetroleum-based hydraulic fluid and submersible drive enclosures. Pin racks are available with nonlubricated components and nickel-coated or nonmetallic cogwheels.

CHAIN-DRIVEN RECIPROCATING RAKE

This reciprocating rake bar screen utilizes a chain-driven rake mechanism to clean the stationary bar rack. The rake arms are attached to

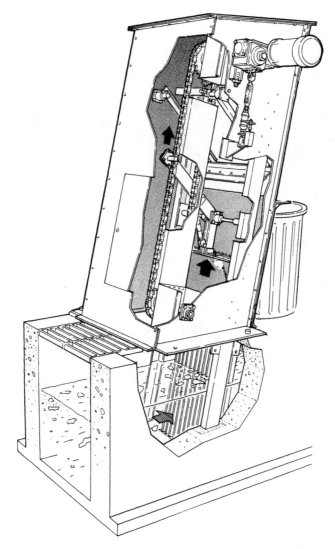

Chain-driven reciprocating rake screen, courtesy of Schloss Engineered Equipment.

a chain-mounted carriage that is raised and lowered by means of two endless strands of roller chain operating over head and footsprockets. Single chain versions of this screen are also available.

Each end of the rake is equipped with rake linkage arms and rollers or sliding blocks that travel in guide channels to maintain the pivot point of

Chain-driven reciprocating rake screen, courtesy of Schloss Engineered Equipment.

Chain-driven reciprocating rake screen, courtesy of Arlat, Inc.

the rake arm to control the engagement/disengagement of the rake with the bar rack. The rake-to-chain connection is equipped with a spring-actuated overload device that allows the rake to pass over obstructions that it may encounter.

The screen chain may be furnished with a chain pitch of 1-1/4″ to 6″ (32 to 152 mm). Chain-driven rake screens have gear motor drive units located in a fixed position on the headframe. The chain-driven reciprocating rake bar screen is generally considered a lighter duty screen than the cog-wheel/pin rack unit and is usually limited to water depths of 8 ft (2.4 m). Front and back cleaned units are available.

CHAIN LIFT RECIPROCATING RAKE

Operation of the cleaning rake may be accomplished through the use of two matched strands of a 4″ (102 mm) pitch roller chain operating in guide tracks. The lower end of each strand of chain is attached directly to the rake arms, and the upper end is fixed to the screen headframe. The raising and lowering of the rake carriage is controlled by the operation of the roller chain over sprockets mounted on a drive shaft in the screen head-frame.

The chain/rake guide tracks are constructed in a rigid frame assembly, supported by the screen drive shaft, which allows the entire guide frame to be pivoted to engage or disengage the rake teeth with the bar rack. This pivoting movement is controlled by an electric actuator with a fractional horsepower drive motor that "powers" the rake into position.

A rotary limit switch on the screen drive shaft limits the up and down movement of the rake, and a cam-type limit switch on the actuator drive controls the rake engagement and disengagement.

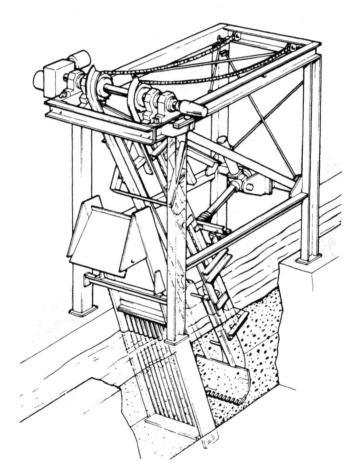

Chain lift reciprocating rake screen, courtesy of Brackett Green.

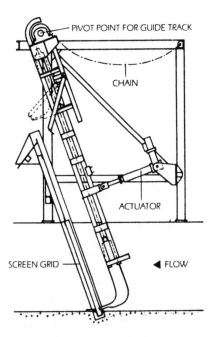

Chain lift reciprocating rake screen, courtesy of Brackett Green.

The screen is protected against rake overloads through the use of current sensing devices in the circuits of the screen drive and actuator motors. If the rake encounters an obstruction in any mode of operation (e.g., rake engage/disengage or rake ascent/descent), it is immediately recognized by the sensors as an increase in amperage, and the rake begins to retract from the bar rack. As the rake clears the obstruction, the subsequent lowering of current allows the rake to continue its operation at the point that it was interrupted. The rake's ability to sense an obstruction and continue to operate it is referred to as "profiling."

Excess chain that accumulates during the raising of the rake is stored on chain collecting bars located within the screen headframe. The maximum headroom required for this screen is approximately equal to the rake discharge height plus 6'-6" (2 m).

CABLE-OPERATED RECIPROCATING RAKE

The cable-operated reciprocating rake bar screen is one of the oldest types of mechanically cleaned screens. These screens use a carriage-

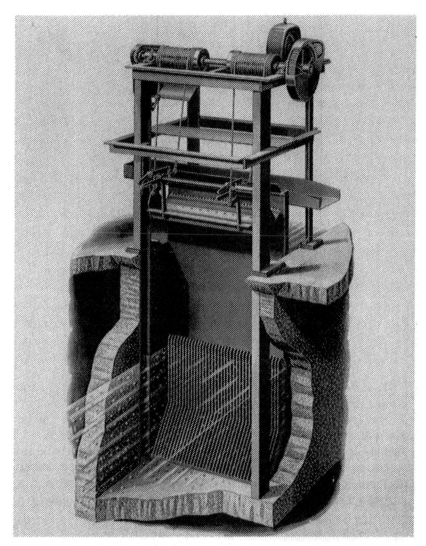

Cable-operated reciprocating rake screen, courtesy of FMC Corp.

mounted rake assembly that is raised and lowered by one, two, or four stainless steel cables, or "ropes," to clean a stationary bar rack. The cables operate from grooved drums mounted on a common headshaft. Rake travel drums are keyed to the shaft to control the up and down movement of the rake, and rake tooth positioning drums are bronze bushed to operate freely. Controlled slippage between the drums will result in the pivoting of the rake teeth.

The rake carriage is equipped with guide blocks that travel vertically within channel guides embedded in the walls of the concrete channel to assure positive rake engagement. A pivoting rake wiper may be mounted on the screen headframe or furnished as an integral part of the rake carriage mechanism.

Cable-operated screens are available in widths of up to 20 ft (6 m).

HYDRAULICALLY OPERATED RECIPROCATING RAKE

The raising/lowering and engaging/disengaging motions of a reciprocating rake screen can be provided through the use of hydraulic cylinders. This screen design is generally considered for light-duty applications with channels less than 6ft (1.8 m) deep and 4 ft (1.2 m) wide. The use of a cable-type lift cylinder or an auxiliary debris conveyor may allow use on deeper installations.

The cleaning stroke begins as a hydraulic cylinder mounted between the headframe and rake arm and is retracted to pivot the cleaning rake into the bar rack. A second hydraulic cylinder mounted at the top of the screen headframe is used to raise and lower the cleaning rake. This cylinder may be a rod or cable type.

The synchronized motion of the cylinders is powered by a small hydraulic drive unit. Pressure relief valves provide overload protection to prevent equipment damage.

SCREW-OPERATED RECIPROCATING RAKE

This screw-operated bar screen is a reciprocating rake bar screen that has been used in channels with water depths of up to 10 ft (3 m).

The up and down rake movement is provided by a ball bearing nut that travels the length of a rotating screw. The vertically positioned screw is directly coupled to a reversing drive motor and operates in stationary upper and lower bearings mounted to the screen headframe. The rake carriage is fitted with a ball bearing drive nut that traverses the length of the screw as it is rotated.

Rake arms project down from the rake carriage and are equipped with upper and lower guide rollers. The rollers travel in guide channels to provide tooth engagement on the rake's ascending stroke and disengagement on the descending stroke. Upper and lower proximity switches limit the travel of the rake carriage.

Hydraulically operated reciprocating rake screen, courtesy of Franklin Miller.

126

Hydraulically operated reciprocating rake screen, courtesy of *Zenith*.

Screw-operated reciprocating rake screen, courtesy of Enviro-Care.

MULTI-RAKED BAR SCREENS

Multi-rake bar screens use multiple, chain-mounted rakes to clean a single, stationary bar rack. The primary advantage of multi-rake bar screens is their ability to continually remove large volumes of debris from the face of the bar rack. The time that it takes to clean the bar rack is significantly reduced through the use of multiple rakes. Based on a typical rake speed of 10 ft (3 m) per minute and a typical rake spacing of 8 ft (2.4 m), a rake will pass over a given portion of the bar rack every forty-eight seconds. This short cycle time is important with deep channels or those applications that may experience periods of heavy debris loading.

The major disadvantage of the multi-rake designs is the quantity and location of many of the mechanical components. These designs typically use a series of rakes that are mounted on two endless strands of chain, operating over head- and footsprockets. A portion of the chain (and usually other mechanical components including sprockets, bushings, and shafting) operates underwater. This increases the difficulty of inspection and maintenance.

Power is usually provided by an electric motor driving a helical or worm gear speed reducer. Screens are available with shaft-mounted drive units directly mounted on the screen headshaft or with a roller chain drive operating over a drive/driven sprocket arrangement. Rake travel speeds range between 7 and 15 ft (2.1 and 4.5 m) per minute.

Cleaning rakes should be furnished in multiples of two and should be spaced at equal intervals ranging from 2 to 12 ft (0.6 to 3.7 m).

Multi-raked bar screens, courtesy of Hubert Stavoren BV.

FRONT CLEANED, FRONT RETURN BAR SCREEN

Front cleaned, front return bar screens use an "endless" series of chain-mounted rakes to clean an inclined bar rack. This model bar screen has been used extensively on standard-duty sanitary sewage flows.

Both the ascending (cleaning) and descending (return) runs of chain are located on the front, upstream side of the bar rack. Front cleaning provides positive cleaning of the front and sides of the bars by the toothed rakes, and front chain return eliminates the possibility of debris carryover to the downstream side of the screen.

Front cleaned multi-rake bar screen, courtesy of Envirex, Inc.

Front cleaned multi-rake bar screen, courtesy of Amwell.

Front cleaned bar screens pivoted for maintenance, courtesy of Envirex, Inc.

131

The screen chain is generally a 6″ (152 mm) pitch pintle type and may be of cast iron, steel, or stainless steel material. A polyurethane-coated steel chain is also available.

Head- and footsprockets may be of cast iron or steel with hardened teeth, of nylon, or of ultra-high molecular weight polyethylene construction. Split-type headsprockets are recommended to provide ease of assembly and disassembly.

The screen headframe is fabricated from 3/16″ (5 mm) minimum thickness steel, and the sideframes below the floor level are fabricated of steel having a minimum thickness of 5/16″ (8 mm). The sideframes are formed to provide u-shaped guides for the ascending and descending chains and include shrouds around the footsprockets to prevent debris from becoming caught between the chain and sprocket teeth. A curved bootplate, recessed in the channel bottom, provides a bottom seal and a smooth transition as the descending rakes revolve around the footsprockets and begin their ascent.

Debris is removed on the downstream side of the screen by a pivoting rake wiper mechanism. The mechanism should be equipped with adjustable shock absorbers to regulate the return of the wiper.

The front cleaned, front return bar screen is usually installed at an inclination of six to thirty degrees from vertical, in the direction of the flow. The channel may include guide slots cast into the concrete walls that are arranged to receive the screen framework. The recessed frame will allow for the bar rack's complete use of the channel width.

An alternative method of installation is to mount the screen's headframe on pillow block bearings located above the operating floor. The entire screen may then be pivoted out of the channel for maintenance or inspection.

Combination Bar/Grit Screen

One variation of the front cleaned, front return bar screen has been designed to remove both screenings and grit in a single unit. This screen can eliminate the need for separate screen and grit removal systems in some small treatment plants.

This screen utilizes an adjustable baffle and a steeply inclined grit sump to promote settling of grit at the base of the screen. The cleaning rakes are fitted with perforated buckets that collect grit from the sump as they revolve around the footshaft. A "knocker" aids in removing grit from the bucket, and a pivoting wiper mechanism cleans the rake teeth of any material that is not gravity discharged.

This screen is inclined at thirty degrees and is available in widths of 2 to 4 ft (0.6 to 1.2 m).

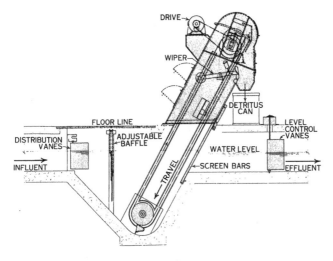

Combination bar/grit screen, courtesy of FMC Corp.

FRONT CLEANED, REAR RETURN BAR SCREENS

The front cleaned, rear return bar screen is considered a "heavy-duty" screen that can be used in applications with heavy solids loading, such as those encountered in combined sewer overflows (CSO). It is also preferred over front return screens for installations with deep and/or wide channels. It consists of multiple rakes mounted on two strands of chain that ascend to clean the front side of the bar rack and return on the rear, downstream side of the bar rack.

A pivoting deflector plate is required at the bottom of the screen to prevent debris from jamming in the boot area as the rakes revolve around the footsprockets. The deflector plate is pivoted out of the way by each ascending rake and closes by gravity as the rake passes.

This bar screen is installed at inclinations of six to fifteen degrees. If a recessed boot is provided or if large amounts of sand or grit are expected, special bucket-type rakes should be interspersed with the standard cleaning rakes to remove the settleable material.

Discharge is accomplished on the screen's rear side after the rakes rotate over the headshaft. To assist in debris removal and prevent debris from falling into the downstream side of the channel, an idler shaft or chain guide track is used to locate the discharge point directly over the screenings receptacle.

The structural frame components above the operating floor should be fabricated of steel having a 1/4 " (6 mm) minimum thickness, and the frame

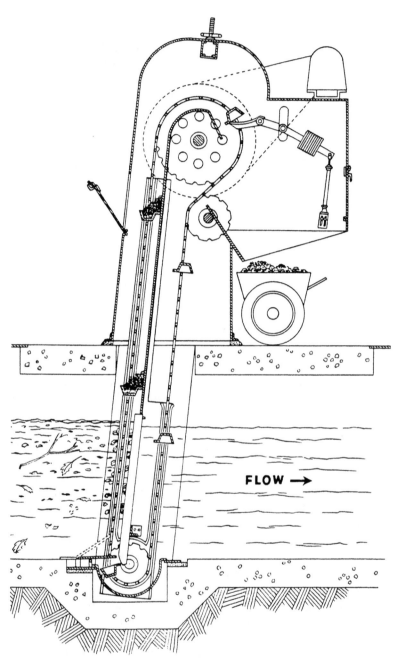

FLOW →

Front cleaned, rear return bar screen, courtesy of Envirex, Inc.

below the operating floor should have a minimum thickness of 3/8″ (9 mm).

The rake chain is usually a pintle or combination type with a pitch of approximately 6″ (152 mm).

Fine Screen

Several manufacturers offer a front cleaned, rear return bar screen equipped with continuous cleaning rake elements. These elements may be of stainless steel or high-impact plastic construction and are designed to automatically "eject" debris as they rotate around the screen headsprockets. These in-channel screens are further discussed in Chapter 7.

BACK CLEANED BAR SCREENS

Back cleaned bar screens are available in rake widths from 2 to 10 ft (0.6 to 3 m) and for applications with water depths generally less than 12 ft (3.7 m). The raking mechanism of this screen is located on the downstream (back) side of the bar rack, with only the rake teeth, or tines, projecting through the bars in the cleaning position. This arrangement has the advan-

Multiple rake, back cleaned bar screen.

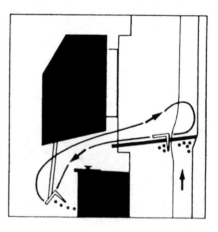

Back cleaned reciprocating rake bar screen, courtesy of Schreiber Corp.

136

tage of minimizing the possibility of debris jamming the rake mechanism or wedging between the bar rack.

The rakes are chain-mounted and revolve over head- and footsprockets. As the descending chains revolve around the footsprockets, the rake teeth engage the bar rack from the downstream side and project into the flow by 4″ to 7″ (102 to 178 mm).

The bars are anchored at the channel bottom and are supported and spaced by the ascending rakes. The rakes, "combing" action through the bar prevents the use of fixed horizontal bar supports at intermediate levels and limits the effective depth of the screen to approximately 12 ft (3.7 m).

Vertical Back Cleaned Screens

Vertical back cleaned bar screens require very little floor space and channel area and are frequently used in high-lift pump stations and deep sewers where there are considerable differences between the operating floor and channel invert elevations. This screen has a maximum effective bar rack depth of approximately 12 ft (3.7 m) but may provide a vertical lift of up to 60 ft (18 m).

The rakes, spaced at 6- to 8-ft (1.8- to 2.4-m) intervals, also serve as horizontal bar spacers. Most screens have a pivoting spacer mechanism located near the top of the bar rack that helps maintain proper bar spacing. The bar spacer is pivoted out of position whenever ascending rakes pass.

Screenings are removed on the front, upstream side of the screen by an automatic rake wiper. A pivoting deadplate and chute are required to assure that screenings do not fall back into the channel.

Reduced weight, cast polyurethane rake assemblies are available and recommended for these screens.

Inclined Back Cleaned Screens

Inclined back cleaned screens are furnished with the bar rack positioned at an angle of nine to thirty degrees from vertical. Unlike vertical back cleaned screens, these units discharge screenings on the downstream side of the screen.

This screen is generally considered a "heavier duty" design than vertical back cleaned bar screens because its inclination allows it to handle more debris.

Some designs are equipped with bars that extend from the channel bottom, up the front of the screen, and over the headshaft to the discharge position. This design helps facilitate debris removal and reduce carryover by stripping debris from the rakes.

These screens are furnished with a self-supporting frame and include

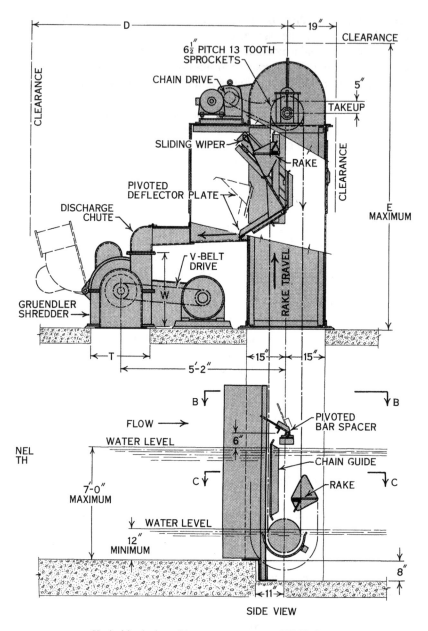

Vertical back cleaned bar screen, courtesy of FMC Corp.

guide tracks for the ascending and descending chains. Bar racks may be fabricated of rectangular or round bars.

CATENARY BAR SCREENS

Catenary bar screens were developed for unattended use at large storm water pumping and drainage projects where the debris loads are heavy or varied, and screen use may be infrequent. Catenary bar screens have also been applied as primary screens at wastewater treatment plants.

Like other multi-rake screens, the catenary bar screen utilizes a series of chain-mounted toothed rakes to clean a stationary bar rack. The most significant difference is that the Catenary Screen has no footshaft or lower chain guide assembly. The screen chain descends into the channel to form a "catenary" curve near the channel bottom.

Catenary bar screen, courtesy of E&I Corp.

Dual shaft catenary bar screen, courtesy of E&I Corp.

The underwater path of the chain and rakes is unguided, allowing them to ride over an accumulation of debris without jamming. The rakes are weighted to bear down against the face of an inclined bar rack and collect and retain debris as they ascend to a discharge point above the operating floor. Rake cleaning is accomplished by a pivoting rake wiper mechanism that scrapes the screenings from the rake into a suitable receptacle.

Combination or pintle chain with a 6″ (152 mm) pitch is most frequently used for these screens.

The ends of each rake may be equipped with a replaceable, hardened steel or UHMW polyethylene wear pad. This pad engages the bars and/or deadplate on the rake's ascent and minimizes contact and subsequent wear of the rake teeth and screen bars.

Catenary bar screens are available in widths ranging from 2 to 30 ft (0.6 to 9.1 m) and depths to 50 ft (15.2 m) or more. Bar rack inclinations of fifteen to forty-five degrees are recommended.

The catenary loop of the chain may prevent the bottom 4″ to 10″ (102 to 254 mm) of the bar rack from being effectively cleaned. It is recommended that the lowest portion of the bar rack be curved into the flow to prevent debris accumulations at the channel bottom.

Units are available with single or double shaft designs. Double shaft screens are recommended for heavier duty application and those installations with bar rack inclinations of thirty degrees or more.

The structural support for a catenary screen drive mechanism consists of a relatively simple headframe assembly that is anchored to the operating

Dual shaft catenary bar screen, courtesy of E&I Corp.

Counter-weighted catenary screen rakes, courtesy of Fairfield Service Co.

deck. The descending, return run of the chain is guided by runway angles bolted to the channel walls that may extend to within 3 ft (0.9 m) from the channel bottom.

OTHER BAR SCREENING EQUIPMENT

ARC SCREENS

An arc screen is a self-cleaning bar screen consisting of a curved bar rack that is cleaned by one or more revolving rake assemblies. These screens are designed for applications in small plants with channel depths usually less than 7'-0" (2.1 m), and they are available with a variety of rake configurations.

The raking mechanism is mounted on a drive shaft that operates in pillow block bearings and slowly rotates about a horizontal axis. The rotation of the cleaning rake cleans the radius of the stationary bar rack and elevates debris to a discharge point at the top of the rack, where it is cleaned by a wiper mechanism. Some units are provided with a pivoting debris plate to prevent debris from falling back into the channel.

The rake teeth engage the bar rack and may be furnished with an optional spring-loaded mechanism to override minor obstacles. The bar rack is constructed of steel bars formed in a radius that is approximately equal to the sum of the channel depth and the rake discharge height.

One variation of the arc screen utilizes bristle brushes in place of toothed rakes, and the debris is swept to the discharge point. This screen is further described in Chapter 7.

Arc screens may be furnished with direct coupled, chain and sprocket, v-belt, or shaft-mounted drive units. Overload protection can be provided by a torque-limiting coupling, shear pin device, current sensor, or torque overload switch.

Full Rotary Arc Screens

The cleaning stroke of the full rotary arc screen is a complete 360 degree circle, and its cleaning mechanism is very simple and reliable. Screens may be fitted with one, two, or four cleaning rakes.

Arc screen, courtesy of Kruger.

Full rotary arc screen, courtesy of Longwood Engineering.

Full rotary arc screen, courtesy of COS.M.E.

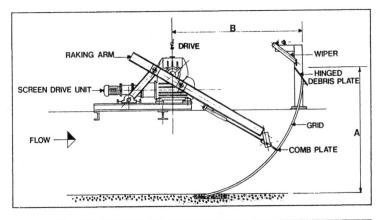

'A' Channel Invert to discharge point (M)	1000	1250	1500	1750	2000	2250	2500
'B' Minimum length of channel (M)	2000	2300	2750	3100	3500	3800	4100

Semi-rotary arc screen, courtesy of Brackett Green.

Screens may be installed in concrete channels for through-flow operation or above floor openings and drains for down-flow applications. Screens up to 4 ft (1.2 m) wide may be installed in concrete channels, and widths up to 12 ft (3.7 m) are available for down-flow screens having 3 to 6 ft (0.9 to 1.8 m) rake diameters.

Semi-Rotary Arc Screens

The rotary motion of the screen drive unit can be converted into a semi-rotary action by means of a simple pivoting linkage or a hydraulic actuator. As the rake completes its upward cleaning stroke, it disengages and returns to the bottom of the curved bar rack. It then re-engages the bars and repeats the cycle.

The semi-rotary arc screen is furnished with a single cleaning rake. Its primary advantage is that it requires approximately one-half of the channel length and headroom required by a full rotary screen.

SOCK SCREENS

Although not a true "bar screen," the sock screen may be effective in some low-flow bar screening applications. This screen utilizes a mesh

Sock screens, courtesy of Hydro-Aerobics, Inc.

"sock" fitted to a prefabricated steel frame. Water flows out an opening in the frame and through the mesh sock. Solids are retained on the inside of the sock until it is removed and replaced or cleaned.

When screenings contain fecal matter or other organics, the socks may be fitted with an agitation mechanism that helps break up the organics so they can pass through the sock. The mechanism consists of a reciprocating arm attached to the closed end of the sock. When activated, the motorized arm agitates the sock and "washes" the accumulated solids.

Based on application requirements, mesh fabric may be furnished with openings that range from 1/8″ to 1″ (3 to 25 mm). Sock lengths of up to 6.5 ft (2 m) or more offers sufficient screening area to provide extended operation in many applications.

MANUALLY CLEANED BAR RACKS

Manually cleaned bar racks are frequently the only type of screening employed in small wastewater treatment plants. Larger plants that use mechanically cleaned bar screens for primary screening purposes may also use manually cleaned bar racks in overflow bypass channels.

Manually cleaned bar racks should be limited in depth to that which can be conveniently raked by hand. The racks are normally placed at an inclination of thirty to forty-five degrees from horizontal to increase available screen area by 40 to 100%. This inclination will also assist the operator in his ability to drag up large debris with the manual rake.

Bar spacings for manual rakes are typically 1″ to 2″ (25 to 50 mm), and the size of a manually cleaned rack usually measures approximately 20 ft² (1.9 m²) of screening area per million gallons per day of treatment capacity. A horizontal, perforated drainage plate may be furnished at the top of the rack to temporarily store debris for drainage.

Manually cleaning may be required two to five times per day. Debris buildup on the racks must be carefully monitored to insure reasonably free flow into the treatment plant.

STORM WATER OVERFLOW SCREENS

Many of the previously described bar screen designs have been successfully employed on combined storm overflow (CSO) systems. In most cases, these screens are identical to those used in other applications. However, several bar screens have been specifically developed to remove floatables from storm water overflows and return the solids to the sanitary sewer underflow.

One such screen consists of a rectangular screen grid, with openings of 1/4″ (6 mm), installed in an interceptor splitter box. When storm water flow exceeds the capacity of the sanitary sewer system, the overflow is directed through the screen before being discharged to the receiving water body. Solids that accumulate on the screen are removed by a reciprocating carriage and carried away with the sanitary sewer underflow. Operation is controlled by a level switch, and the hydraulically driven cleaning mechanism is located on the downstream side of the bar grid.

Another CSO screening arrangement utilizes a variation of the arc screen to remove floatables from storm water overflow. This curved bar grid is installed at the overflow weir level in a splitter box or stilling bay and prevents passage of solids with the excess flow. A rotating rake arm

Storm water overflow arc screen, courtesy of Longwood Engineering.

cleans the bar grid and returns solids to the underflow. This screen is furnished with bar spacings ranging from 1/2 " to 1 " (13 to 25 mm).

BAR SCREEN COMPONENTS

BAR RACK

The common component of all bar screens, whether mechanically or manually cleaned, is the bar rack. The bar rack is a stationary grate constructed of equally spaced vertical bars. Uniform bar spacing is to be maintained through the use of horizontal support members with milled slots or welded spacers to which the vertical bars are attached.

Bar racks typically extend from the channel bottom to a point at least 9 " (229 mm) above the maximum water lever, where they are attached to a steel deadplate that extends to the debris discharge point. Many screens are designed to have their framework recessed in the concrete walls of the channel so that the entire width of the channel is available as an effective screening area.

The bar rack should be structurally sound and capable of withstanding the loads imposed by a minimum headloss of 2 ft (0.6 m). Although many sizes and shapes of bars may be utilized, it is normally recommended that the section modulus be equal to or greater than that of a rectangular bar measuring 3/8″ (9 mm) thick by 2″ (50 mm) wide. The width, depth, bar spacing, and headloss requirements will determine the actual size requirements.

Debris often becomes wedged between rectangular bars, just beyond the reach of a bar screen's rake teeth. Most manufacturers are now offering

Reciprocating carriage storm water overflow screen, courtesy of Jones & Attwood.

Storm water overflow screen, courtesy of John Meunier Corp.

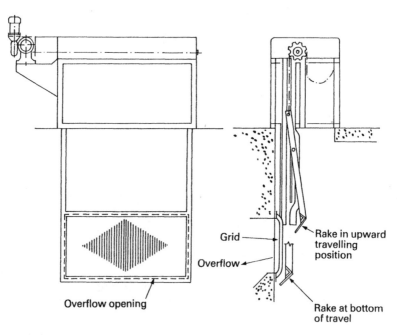

Vertical storm water overflow screen arrangement, courtesy of Longwood Engineering.

racks constructed of bars having a trapezoidal cross section to help avoid this problem. Trapezoidal bars are oriented on the rack using the base of the trapezoid as the upstream screening surface, with the bar thickness decreasing progressively in the direction of the flow. This prevents debris from becoming wedged between the bars; the debris either remains matted on the face of the rack or flows completely through it.

When specifying trapezoidal bars, it should be considered that some bar sizes may be nonstandard and may not be stocked by all steel supply houses. Most manufacturers standardize and stock one or two bar sizes that they have purchased in large, mill-run quantities. If a particular cross section is specified, it may provide a significant price advantage to the manufacturer that stocks the specified size. If trapezoidal bars are preferred, it is recommended that the specification reference be an approximate bar size and/or a minimum acceptable section modulus.

Another variation of the trapezoidal bar is a design with a "teardrop"-shaped cross section. This bar offers the advantages of the trapezoidal design, as well as hydrodynamic advantages in applications with fine openings.

At least one manufacturer offers a bar rack constructed of rectangular bars positioned at an angle to the flow. This design is said to result in improved flow through the unit and, when used in conjunction with a sawtoothed rake design, provides full engagement of the cleaning rakes with the bars.

When possible, the bar rack should be constructed so that it can be readily removed without disassembling the screen or removing it from the well. Some installations have been provided with a bar rack that can be automatically lifted from the channel during extreme flow or headloss conditions.

Deadplate

In most applications, it is not necessary or economical to extend the bar rack to the discharge point. Most bar racks terminate at some point above the maximum water level and are attached to a deadplate that continues to the point of discharge.

Deadplates are normally constructed of 1/4″ (6 mm) thick steel plate and are fabricated with 6″ (152 mm) high sideplates to prevent screenings from falling off the ends of the rakes. The deadplate should be constructed flat and true, so that the clearance between the deadplate and rake teeth is as small as possible.

A discharge chute or apron may be used to direct the screenings from the apex of the deadplate to the screenings, receptacle or conveyor. The apron is usually of 1/8″ (3 mm) thick carbon or stainless steel construction.

CLEANING RAKES

Mechanically cleaned bar screens are equipped with cleaning rakes that have teeth cut to match the openings of the bar rack. The teeth on front cleaned designs should project at least 1/2 " (13 mm) into the bar rack to effectively clean the front and sides of the bars, and the rakes should provide a minimum lifting shelf of 5-1/2 " (140 mm). The rakes on back cleaned designs should penetrate through the bar rack and provide a minimum lifting surface of 4 " (102 mm) on the upstream side of the screen.

The cleaning rakes on reciprocating rake screen designs should be able to collect and retain all the accumulated debris on the face of the bar rack in a single cleaning cycle. It is recommended that the cleaning rakes on these screen designs have a minimum 8 " (203 mm) wide lifting shelf, although some rakes as wide as 14 " (356 mm) may be used.

Rake teeth are usually flame-cut or machined out of steel plate. Bar spacings of less than 1/2 " (13 mm) usually require that they be machined.

It is recommended that toothed rake segments be considered for use with bar spacings less than 1 " (25 mm). Rake segments are typically 12 " to 24 " (305 to 610 mm) long, and they are bolted directly to the rake body. They may be individually adjusted to minimize tooth/bar rack interference and compensate for tolerances accumulated during fabrication of the bar rack. Cast iron or cast steel rake tooth segments are available from some manufacturers.

Several manufacturers offer rake designs that allow individual rake tooth segments to ride over minor obstacles. The rake tooth segments are mounted to the rake body by means of a spring-loaded hinge. As a small obstacle is encountered, the rake tooth segments pivot down and out of the way without overloading or stopping the screen mechanism.

Nonmetallic rakes and rake segments are also available for use on front and back cleaned bar screens. These toothed rakes are bolted to a steel rake body, resulting in an overall rake weight reduction of approximately 65% per rake assembly. The low coefficient of friction and the abrasion resistance of the urethane or polyurethane material also provide increased performance.

The screen mechanism and rakes are typically designed to provide a minimum lifting capacity of 60 lb/ft (90 kg/m) of rake width.

SCREEN CONTROLS

The purpose of a screen control system is to hold the differential headloss and velocity across the screen at or near the optimum "clean screen" levels. Although some screens may be able to be operated manually, most installations require some type of automated controls to insure

Bar screen control panel, courtesy of Arlat.

reliability and economy of operation. There are few limits to the versatility that can be designed into a screen control system.

Hand-Off-Auto

A simple screen control system may consist of a Hand-Off-Automatic (HOA) switch. In the "Hand," or manual, position, the screen will continue to operate until turned to the "Off" position. When the screen is operated manually, the optional automatic control devices such as the time clock, limit switch, float switch, or differential level indicator have no effect.

In the "Automatic" position, the screen operation will be initiated by one of the optional control devices mentioned above.

Float Switch

A float switch may be used to start and stop the screen operation based on water level in the channel. Additional float switches can be located at various elevations to operate annunciators, pumps, or other equipment as a result of changing water levels.

Differential Level Controller

A differential controller may be used to monitor the water level on the upstream and downstream sides of a bar rack and automatically start and stop the screen based on the differential water level. Ultrasonic and pneumatic systems are available to measure the water levels.

As debris accumulates on the screen, the downstream water level begins to fall below the upstream level. When this differential water level increases to a predetermined amount, the screen mechanism is activated.

Screen control systems should be designed to allow the screen to complete its operating cycle after the headloss condition has been relieved. Controllers for multi-rake bar screens should be furnished with an adjustable timer that will insure that the screen continues to operate for 1-1/3 revolutions after the leadloss is relieved to prevent excess screenings from drying out on the rakes.

A programmable timer can be used in conjunction with a differential controller. The timer can be set to activate the screen after a predetermined time, whether or not a differential headloss has occurred. Should a headloss occur, the differential controller will override the timer.

ENCLOSURES AND HOUSINGS

Steel, stainless steel, fiberglass, and aluminum enclosures are available to protect screen drive machinery, assist in controlling odors, and increase operator safety. Screens that use a spray system to aid in screening discharge require an enclosure to contain overspray. Depending on location and screen type, a bar screen enclosure may range from an expanded metal safety enclosure to a heated, insulated, and airtight housing. Enclosure designs should provide access to all of the screen's rotating machinery for maintenance purposes.

BAR SCREEN SELECTION AND SIZING PROCEDURES

Proper bar screen selection and sizing will insure satisfactory mechanical and process performance and increase the efficiency of downstream equipment and processes in the treatment plant. Criteria used to determine the type of screen best suited for a particular application is similar to that used in other treatment plant processes:

- particle size and volume to be removed
- diurnal flow variations and influent characteristics
- maximum and minimum water levels
- plant hydraulics and allowable headloss
- qualifications and availability of operators and maintenance personnel

- engineers/operators preferences
- final disposition of screenings

The procedure most frequently used to establish the size of a bar screen involves the determination of the size of the bar rack, plus suitable freeboard, to pass the required flow at a velocity that prevents sedimentation. After the size of the bar rack has been established and the screening volume has been estimated, a suitable rake mechanism can be selected.

Design constraints such as channel width, depth, and headroom may limit the choice of screen models available for a particular application. But with the many types of bar screens available, it is usually possible to select from among more than one type of screen design.

As with most types of mechanical equipment, maintenance requirements and reliability are important aspects of design. Some applications may require a screen to operate as infrequently as ten minutes per hour, while others may require continuous operation. When operation is necessary, the bar screen should be expected to perform efficiently and effectively, with a minimum of operator attention.

It is suggested that operators of similar bar screen installations be contacted to review the operational history of the preferred models prior to making a final screen selection. This is especially important in the case of large screens or those locations where severe operating conditions are anticipated.

NUMBER OF SCREENS

The number of bar screens required is usually a matter of the engineer's preference. However, it is generally recommended that a minimum of two screens be installed. For smaller treatment plants, the second screen may be a simple, manually cleaned bar rack used during emergencies and installed in an overflow bypass channel.

On the basis of operating and maintenance considerations, two narrow bar screens are recommended over a single wide screen. When two or more screens are used, the screen channel and screening equipment should be designed so that one screen can be taken out of service without adversely affecting the operation of the remaining screens.

When possible, all screens at a single installation should be of the same model, manufacturer, and size. This will insure interchangeability of mechanical components and reduce the required spare parts inventory.

Future screening requirements should always be considered. If it is anticipated that additional screens will be required on future plant expansions, the size and location of the additional screens should be considered. It is usually cost-effective to complete the civil work for a future plant addition during initial plant construction.

BAR SPACING

The bar screen sizing process almost always begins with the screen bar rack and the determination of the particle size that must be removed. The selection of the bar spacing is usually made on the basis of protecting downstream equipment and treatment processes, and a 3/4″ (19 mm) clear opening is adequate for most new treatment plants. However, many wastewater plants prefer finer bar spacings for process or other operational and maintenance reasons.

For some plants, the increased "efficiency" of a finer opening may not justify the increased handling/disposal cost of the increased volume of screenings. Very close spacing of bars may also result in the removal of organic particles necessary to sustain downstream biological treatment processes.

VELOCITY

The size of the channel is determined by a review of the flow conditions. Generally, the channel should be designed to provide a minimum approach velocity of 1.3 fps (0.4 m/s) to avoid sedimentation. Where large amounts of storm waters are to be handled, an approach velocity of approximately 3 fps (0.9 m/s) should prevent grit from settling at the bottom of the bar screen.

The size of the actual screening area required for a bar screen is based on flow, bar rack efficiency, and the desired velocity through the bar rack. In most wastewater applications, it is desirable to maintain a velocity through a clean bar rack of approximately 2 fps (0.6 m/s) during periods of average daily flow. A maximum 3 fps (0.9 m/s) velocity through the bars is commonly used for periods of peak instantaneous flows.

The velocity through the screen bars can be calculated according to the following formula:

$$A = (Q) \ (1.547)/(V)(E)$$

where

Q = flow (mgd)
V = velocity (fps)
E = bar rack efficiency (%)
E = (opening)/(bar size + opening)

After determining the actual screening area required, the channel width and depth can be selected. For example, the screen are determined by the above formula may require 60 ft² (5.6 m²) of actual screen area. This could be configured in a bar rack measuring 6 ft (1.8 m) wide and 10 ft (3 m) deep or 3 ft (0.9 m) wide and 20 ft (6 m) deep. It is recommended that sufficient freeboard be added to the maximum water level to allow for flow surges.

A headloss develops across the screen as debris collects on the face of the bar rack. This headloss should be relieved as the screen is operated and collected debris is removed. The theoretical headloss can be computed according to the following formula:

$$H = k(V^2 - v^2)2g$$

where

H = headloss
V = velocity through bars
v = approach velocity
g = acceleration (32.2 ft/sec²)
k = friction coefficient (1.43, typical)

CHANNEL DESIGN

Mechanically cleaned bar screens are usually installed side-by-side in concrete channels. Single screen installations will require a bypass channel with a manually cleaned bar rack for emergency operation. Channel designs should incorporate the use of stop logs or sluice gates to isolate each screen and facilitate dewatering for maintenance and inspection purposes.

The channel area upstream of the screen should be straight for several feet to insure that the flow is uniformly distributed on the face of the bar rack. The channel floor and sidewalls should be smooth and free of projections that might otherwise prevent uniform velocity distribution.

There are several methods of installing bar screens within a channel. Screens may be supported at the operating deck or channel invert. They may be inserted in recesses cast in the concrete channel walls, flush mounted to the walls and grouted in place, or installed within vertical steel guide slots. The arrangements available are dependent on the screen design selected, and screen manufacturers should be consulted for specific details.

Most screen channels are designed to be as shallow as practical to minimize the headloss through the treatment plant. The location of pumps, grit chambers, Parshall flumes, wet well, etc., downstream of the screen will affect the actual water depth in the channel. In some applications, it may be necessary to install stop planks, acting as weirs on the downstream side of the screen, to increase the sewage depth.

Typical minimum and maximum channel widths for mechanically cleaned bar screens are 2′-0″ and 14′-0″ (0.6 and 4.3 m), respectively.

SCREENINGS VOLUME

The volume of screenings removed by a bar screen varies greatly, depending on the clear opening between bars and the character of the sewage. Normal variations for domestic sewage range from 0.5 to 10 ft³ of screenings per million gallons of screened sewage. The screenings volume will, of course, increase as the width of the openings is decreased.

For most primary wastewater treatment applications, the screenings volume can be expected to more than double for each 1/2″ (13 mm) reduction in bar spacing. For example, a bar screen having a 1/2″ (13 mm) bar spacing will yield an average debris volume of 7.5 ft³ (0.2 m³) per million gallons of screened sewage as compared to 1.25 ft³ (0.03 m³) of debris for a screen with 1-1/2″(38 mm) openings.

Typical wastewater screenings contain approximately 80% moisture and weigh 40 to 60 lb (18 to 27 kg) per cubic foot.

CHAPTER 7

Fine Screens

THE development of improved screening equipment has resulted in the increased use of screens with smaller, "finer" clear openings. Fine screens are generally classified as those screens whose clear openings are less than 1/4" (6 mm) and may range as low as 0.008" (0.2 mm). Units with openings at the lower range of this spectrum are often able to reduce solids levels to primary treatment levels.

Fine screens are used most commonly in wastewater treatment plants in applications that include

- solids removal to protect downstream equipment and processes
- primary treatment, in lieu of primary clarifiers
- solids recovery in industrial process streams
- sludge screening
- scum dewatering
- screenings or grit dewatering

Although a fine screen installation may cost one and a half times more than a conventional bar screen installation, their application is based on the fact that the screened water or sludge is easier to treat, and the operational and maintenance costs of downstream equipment are often reduced.

The higher installation cost may also be offset by the elimination of primary clarification. A fine screen installation usually requires 10% of the land area and normally has a capital cost 30 to 50% less than a comparably sized primary clarifier.

Screen headloss and the volume of solids removed increase exponentially as the size of the screen opening decreases. Therefore, the application of a fine screen necessitates a careful review of the plant hydraulics, the screenings handling system, and the sizing of downstream equipment and processes. This review is especially important in a retrofit installation where a coarse bar screen is being replaced by a fine screen.

The handling and removal of screenings can be an important part of the

159

overall operational success of a fine screen installation. Fine screens typically remove 6 to 32 ft³ of solids per million gallons of wastewater. This is as much as ten times more screenings than a coarse bar screen installation.

Adequate provisions should be made to contain and/or dispose of screenings for health, odor, and aesthetic reasons.

Solids removal attributed to fine screens may result in a reduction of BOD levels by 5 to 25%, TSS by 15 to 30%, grease by 30 to 50%, and up to 90% of all floatables. Some types of fine screens may also pre-aerate plant inflow to a dissolved oxygen level of 2 mg/L or more. Although this may affect the sizing of downstream unit processes, most engineers give fine screens little or no credit for the BOD removal or pre-aeration that they may accomplish. Instead, they focus on the removal of inorganic solids and trash that may affect performance of downstream equipment.

The reduced size of openings in fine screens may require that the screening surfaces be manually cleaned with brushes or steam every six to eight weeks. This may be especially true in applications encountering large amounts of grease in the screen feedwater.

The various types of fine screens may be catagorized and differentiated by the way that they are positioned relative to the incoming flow, the screening medium utilized, or the cleaning mechanism employed.

It is recommended that the equipment design, sizing, and selection be discussed at length with potential equipment manufacturers. Pilot tests may be conducted on "nonstandard" liquid streams and wherever fine screens are being used to replace existing treatment processes.

FILTER SCREEN

Filter screens are in-channel screening devices that consist of an endless cleaning grid, or belt, operating around the headsprockets and footsprockets or lower curved guide rails. The screens are oriented perpendicular to the channel flow. The upstream face of the moving grid collects and retains debris, elevates it out of the flow, and discharges it into a suitable receptacle.

Many filter screens rely on the screenings themselves to enhance solids removal. Solids retained on the screen surface form a tight mat that reduces the effective opening of the screen and contributes to the removal process.

Discharge usually takes place on the back, downstream side of the screen, after the debris-laden screening surface has rounded the headshaft and is beginning its descent.

In-channel filter screen, courtesy of Longwood Engineering.

The fact that these screens can be mounted in a below-grade channel results in the elimination of a pumping step in some new plants, and the screens can be retrofit to fit many existing plants. Filter screens are available with openings that range from 0.04″ to 1/2″ (1 to 13 mm), channel widths of 1 to 15 ft (0.35 to 4.6 m), and depths to 40 ft (12 m) and can be installed at inclinations of five to thirty degrees from vertical.

Power is introduced by an electric motor/gear reducer drive arrangement and is transferred to the headshaft through a roller chain drive system. Some systems utilize shaft-mounted drive units. The drive motor is typically rated at 1/2 to 2 horsepower (0.4 to 1.5 kW) and operates the screen at travel speeds of 7 to 12 fpm (35 to 60 mm/s).

The screen's structural frame is fabricated from a minimum 3/16″ (4.5 mm) thick steel, or stainless steel, and is usually supported at the opening floor (coping) level. Screen assemblies may be mounted on pillow block bearing to allow the entire screen to be pivoted out of the channel for maintenance or inspection. For pivoting designs, side seals of neoprene prevent debris from bypassing the screen frame.

At least one manufacturer offers a "packaged" system, where the filter screen is mounted in an above-grade tank.

CONTINUOUS ELEMENT FILTER SCREEN

This style of filter screen utilizes individual screening elements mounted vertically on a series of horizontal shafts to form the screen grid.

The cleaning element may be constructed of stainless steel or an impact resistant plastic and are shaped to accomplish both coarse and fine screening simultaneously. The elements are fitted to horizontal shafts equipped with guide rollers on each end. The ascending guide rollers operate in u-shaped track guides fabricated as part of the screen frame. Closure plates seal the area between the outer cleaning elements and the edge of the screen frame.

One variation of the filter screen uses two step-shaped sets of thin vertical plates to form the screening grid. One set of plates remains stationary,

Continuous element filter screen shop test, courtesy of Wheelabrator/Wiesemann.

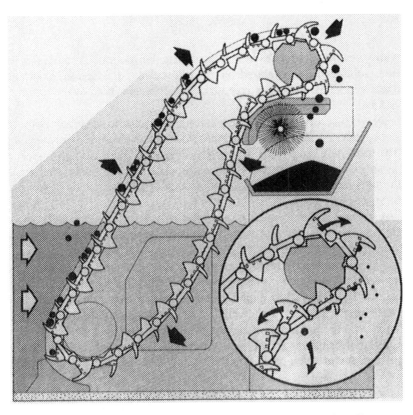

Continuous element filter screen operational diagram, courtesy of Parkson Corp.

Continuous element filter screen, courtesy of Noggerath & Co.

163

while the other set rotates in a vertical, circular motion. Through this motion, solid particles captured on the screen face are lifted up to the next fixed step. This process is repeated until debris is lifted to the top of the channel.

Debris discharge methods vary among the available screen designs. One screen offers a combination drive/unloading headsprocket arrangement. The headshaft is equipped with a drive sprocket made of high-impact resistant plastic. The periphery of the sprockets has flat blades that project through the cleaning elements from inside the screen, dislodging debris as the sprocket rotates.

A second cleaning method uses the rotation of the specially shaped cleaning element to automatically discharge debris as the elements rotate around the headsprockets. The rotation causes a curved portion of the leading elements to move through the spaces of the trailing elements, ejecting debris. The action is then reversed as the belt rotates over a guide rail, causing the elements to wipe themselves clean. Some designs use spray water, rotary brushes, doctor blades, or combinations of these devices to assist with the discharge of stringy or sticky debris.

A variation of this screen is a center flow unit, on which the screening surfaces are oriented parallel to the channel flow. Influent enters the center

Continuous element step screen, courtesy of Hydropress.

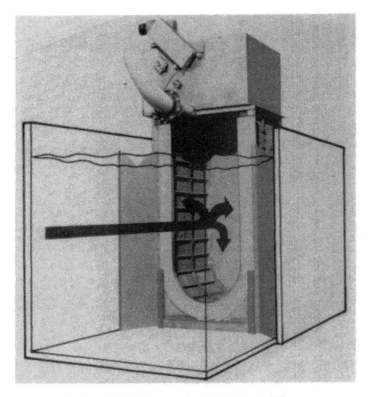

Double entry filter screen, courtesy of Wheelabrator/Wiesemann.

of the screen and exits through both the ascending and descending screening elements. Stationary, self-cleaning grids help reduce screen headloss by funneling solids into the screening chamber.

BELT AND BAND SCREENS

These filter screens are a variation of the traveling water screen and utilize woven wire mesh or perforated plate as a screening medium.

The unit may consist of a continuous belt of stainless steel or plastic wire mesh or a series of interlocking panels fitted with perforated plate or woven mesh inserts. Most designs employ roller chains and sprockets to convey the screening belts/panels.

Increased lifting capacity may be provided by attaching auxiliary lifting shelves to the screening surface at regular intervals. Filter screens equipped with individual screen panels usually rely on the panel design and its orientation to aid in the retention of solids.

Belt screen installation, courtesy of Brackett Green.

Belt screen installation, courtesy of Pro-Ent Corp.

Some belt screens are furnished with a steeply inclined screen approach area that increases screen surface area and minimizes screen headloss. A portion of the screen belt extends upstream at an almost horizontal inclination. The length of this section may correspond to the channel cross section.

Another belt screen design utilizes a series of side-by-side vertical mesh belts as its screening surface. The belts are automatically and uniformly twisted 180 degrees after passing the spray wash system to present a clean surface to the incoming flow. This design may be fitted with an independently operated chain and scraper system to assist in the removal of larger solids.

STATIC SCREENS

Static screens use a stationary, inclined screen deck that acts as a sieve to remove solids from liquids. Static screens are also commonly referred to as "side hill" and "rundown" screens.

Influent enters the screen through a headbox located on the back of the unit. The headbox is appropriately sized and baffled to reduce turbulence and deliver flow to the top of the screen through an overflow weir. Water flowing over the weir accelerates downward and passes over the inclined screen deck. As it cascades over the face of the screen, liquid is stripped away from the solids and falls through the screen openings into a filtrate chamber. Oversized solids continue down the face of the screen and are gravity discharged from its lower edge.

The static screen's most notable features are its compact size, design simplicity, lack of moving parts, noiseless operation, and low operating and capital costs. Most units can be gravity fed and require no power for operation.

The screen decks for these units are typically constructed of wedgewire bars with clear openings of 0.010″ to 0.100″ (0.25 to 2.5 mm). The slotted openings are usually oriented horizontally so that they are positioned

Static screen installation, courtesy of Andritz Sprout-Bauer.

Static screen assembly, courtesy of Hycor Corp.

ninety degrees to the liquid flow. The openings of some wedgewire decks contain slight downward curves, or waves, to help center the flow between vertical supports and minimize blinding.

A combination of the bar design and deck inclination increases the hydraulic efficiency and dewatering ability of the screen. Deck configurations available include straight screens, curved screens, and three-slope designs. Most units are inclined forty-five to sixty degrees, and some are adjustable to suit site-specific operating conditions. Most screen decks are designed to be removable from the screen frame.

Screens are available in widths of 2 to 6 ft (600 to 1800 mm). Hydraulic capacities of the largest, widest units may be more than 1000 gpm (63 L/s). Static screen frames and screen decks are usually manufactured of stainless steel.

Options available include a hinged flow distribution baffle, filtrate viewing port, lower drip lip, back or bottom outlets, back-to-back screen designs for large flows, and removable housings and enclosures.

Some manufacturers also offer static screens with an optional linear driven cleaning brush/spray system. The brush/spray is operated by means

Vibrating-type static screen, courtesy of COS.M.E.

of an electric motor to clean the screen deck. Pressure-fed units are also available for some industrial applications.

VIBRATING SCREENS

A variation of the static screen is the vibrating screen. Vibrating screens are similar to static screen in most respects; however, the screen deck is

Vibrating screen with airtight enclosure and debris conveyor, courtesy of Kopcke.

fabricated as part of a subframe that is mounted on shock absorbers so that vibrations are not transmitted to the main frame of the screen. Vibrating screens are much more effective than static screens in applications that require handling greasy or sticky solids.

ROTARY FINE SCREENS

A rotary fine screen consists of a rotating screen cylinder fitted with a wedgewire or woven mesh screen to remove solids from a liquid stream.

Rotary screens may be internally or externally fed and may be located in a variety of above-grade or in-channel configurations. Screen selection is based on factors that include flow capacity, type of solids to be removed, plant hydraulic profile, and owner preferences.

Most of these self-cleaning screening devices are fabricated of stainless steel throughout and are available with screen openings of 0.010″ to 0.100″ (0.25 to 2.5 mm) or more.

EXTERNALLY FED ROTARY SCREEN

The externally fed rotary screen uses the outside periphery of the cylinder rotating on a horizontal axis as its active screening surface. Influent enters the unit through a baffled headbox that distributes flow evenly along the length of the rotating screen cylinder. Liquid gravity flows through the upper quadrant of the screening surface, into the center of the screen cylinder, and out through the bottom. Solids that are retained on the external surface of the screen are removed by a doctor blade as the cylinder revolves.

The doctor blade is held in contact with the smooth, wedgewire surface

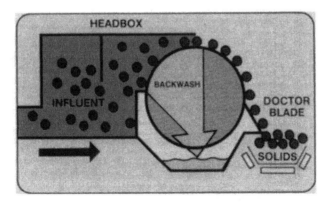

Externally fed rotary fine screen, courtesy of Hycor Corp.

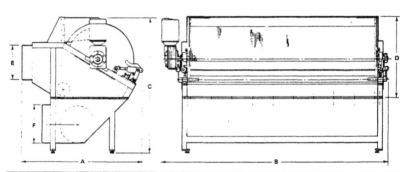

						ALL DIMENSIONS IN INCHES				ALL WEIGHTS IN POUNDS			
Model No.	Motor Hp.	Screen Dia.	Screen Length	Unit Depth A	Unit Length B	Unit Height W/Base C	W/O Base D	Influent O.D. E	Effluent O.D. F	Dry Weight W/Base	W/O Base	Operating Weight W/Base	W/O Base
RSA-2512	½	25	12	45.12	29 09	45 25	30.25	4 50	8.63	300	250	500	375
RSA-2524	½	25	24	45.12	41 59	45 25	30.25	6 63	10.75	600	525	1000	750
RSA-2548	¾	25	48	50 50	66 90	51 38	30.38	10 75	14.00	750	620	1550	1100
RSA-2572	¾	25	72	50 50	90 90	51 38	30.38	12 75	16.00	900	750	2100	1450
RSA-3672	1 2	36	72	70 75	92 50	67 38	43.75	16.00	20.00	1850	1650	5450	3250
RSA-36120	1½-3	36	120	70 75	142 44	66 75	43 75	20 00	18 00[2]	2940	2650	8450	5250

Externally fed rotary fine screen flow diagram, courtesy of Hycor Corp.

172

by a spring-loaded compression assembly and can be furnished with an automatic cleaning mechanism. An optional internal spray backwash system can also be furnished.

The cylinder is supported by shafts on both ends and operates in self-aligning pillow block bearings. A flexible seal located in the headbox rides in contact with the rotating screen surface. End seals prevent debris from bypassing the screening surface.

Internally fed rotary screens are used in general municipal and industrial screening applications and especially where the solids contain sticky or greasy solids or scum. In these applications, the doctor blade is effective in removing solids and minimizing carryover of debris.

Some operating problems are likely to be encountered if the liquid to be screened has large amounts of stringy or fibrous matter. Because flow is introduced parallel to the slot opening, it is possible for the fibrous matter to align itself so that it can partially enter the screen opening and "staple" itself to the screen wires so that it cannot be effectively removed by the doctor blade. Additionally, the same fibrous debris may become stapled to the doctor blade, further reducing the effectiveness of the screen.

Rotary screens are available with drum diameters from 12″ to 48″ (300 to 1200 mm) and lengths from 1 to 12 ft (300 to 3650 mm). Hydraulic capacities range up to 13,000 gpm (820 L/s) per unit. Screens can be located above or below grade, for gravity or pumped feed to the headbox, and a gravity open channel or piped discharge. The 1/2 to 3 horsepower (0.4 to 2.2 kW) screen drive unit is shaft-mounted to operate the unit at a slow rpm relative to the diameter of the cylinder.

INTERNALLY FED ROTARY SCREENS

The internally fed rotary screens utilizes the inside surface of the cylinder rotating on a horizontal axis as its active screening surface.

Influent enters the screen through a trough-type headbox in the center of the cylinder and overflows an adjustable discharge weir. As influent spills onto the side of the rotating screen, the liquid passes through the screen openings, and solids are retained on the inside face of the cylinder. Dewatering of the solids is enhanced by the tumbling action at the bottom of the rotating cylinder, and small diverter plates fixed to the inside of the drum move the solids axially to one end of the cylinder for discharge.

Internally fed rotary screens are well suited for applications where the feedwater has a high fiber content (e.g., pulp and paper industry) and those where solids have a "heavy" consistency (e.g., meat, poultry viscera).

Most internally fed rotary screens utilize wedgewire screens with horizontal slots oriented so that flow is perpendicular to the slot opening. This

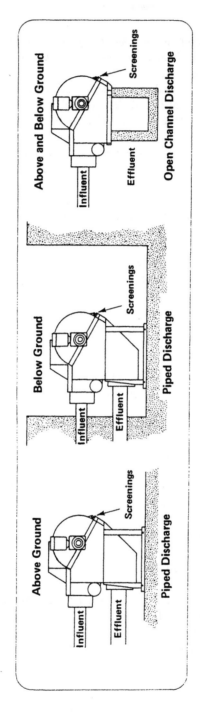

Rotary fine screen layout options, courtesy of Hycor Corp.

Internally fed rotary fine screen installation, courtesy of Andritz Sprout-Bauer.

results in the tendency for solids to form a mat on the screen surface rather than stapling to individual screen wires. As the screen revolves, the mat rolls up the side of the screen until it peels off and falls back toward the bottom of the cylinder so that it can be conveyed to one end for discharge. The perpendicular orientation of the slot opening relative to the liquid flow results in a flat intercept angle and a reduction in the size of the effective opening.

Internally fed rotary fine screen, courtesy of Hycor Corp.

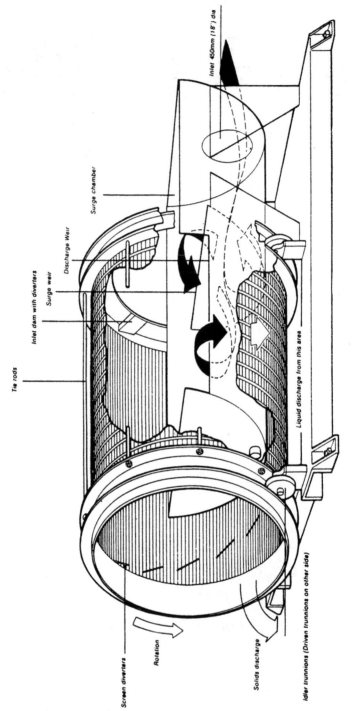

Internally fed rotary fine screen flow diagram, courtesy of Hycor Corp.

Inlet 450mm (18") dia

Surge chamber

Discharge Weir

Surge weir

Inlet dam with diverters

Tie rods

Liquid discharge from this area

Screen diverters

Rotation

Solids discharge

Idler trunnions (Driven trunnions on other side)

Screens may be equipped with an internal or external spray system and/or an external cleaning brush that is used on a continuous or intermittent basis to clean the screen surface and between the slots. Some units are equipped with adjustable diverters to provide operational flexibility.

Internally fed screens are available in diameters ranging from 24″ to 80″ (600 to 2000 mm) and lengths up to 13 ft (4000 mm), to handle flows as high as 13,500 gpm (850 L/s) per unit.

The cylinder is supported at each end by one or more pairs of rollers, or trunnions. A 1/4 to 3 horsepower (0.2 to 2.2 kW) gear motor rotates the drum by means of a power belt, roller chain and sprocket, or spur gear and pinion at approximately 6 rpm.

There are several variations of the internally fed rotary screen described above. Most of these variations are offered for specific screening applications. For example, applications involving higher-than-normal solids loading may utilize an inclined, or a conical-shaped screenings, cylinder to facilitate solids handling. Pressure-fed rotary screens are available for very fine screen openings to 50 microns.

Other variations more closely resemble microscreens or drum screens, which are described in separate sections of this book. Some of these tank-mounted units utilize woven wire mesh screening media and operate with part of the screening surface submerged in the flow. Debris removal is usually accomplished by an external spray system.

HELICAL BASKET SCREEN

The helical basket screen utilizes a cylindrical screening basket mounted in a channel at an angle that may range from thirty to fifty degrees from vertical. Flow enters the lower, open end of the basket and passes through the cylinder, while solids are retained on the interior of the wedgewire or perforated plate screen surface. The solids are then removed from the screen and conveyed up to the discharge point by a helical-type screw conveyor.

Several different cleaning systems are employed to remove solids from the screen. One unit utilizes a toothed rake that revolves around the interior of the screen and rakes debris to the top of the screen basket where it is discharged into a trough by a cleaning comb and conveyed to the top of the channel by a screw conveyor.

Another design utilizes a rotating screen basket with an externally mounted debris spray, while yet another unit utilizes a fixed, semicircular screen basket equipped with a full diameter central screw to clean the bars and convey debris out of the channel.

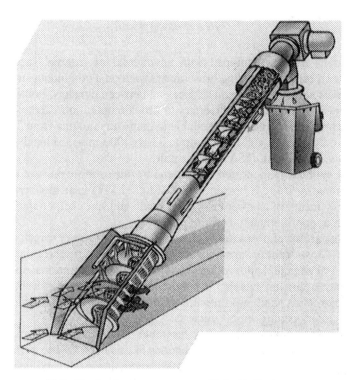

Helical basket fine screen, courtesy of Lakeside Equipment.

Helical basket screen installation, courtesy of Hycor Corp.

178

Multiple screen installation, courtesy of Huber Edelstahl.

Helical basket screens may be equipped with a screening dewatering and compacting system designed as an integral part of the screw conveyor used to auger debris to the discharge elevation. Screens are operated by shaft-mounted gear motors and are available for use in channels ranging from 1 to 6.5 ft (300 to 2000 mm) wide and flows ranging up to 14,000 gpm (880 L/s).

Screw conveyors are available in designs that include a central shaft and those of the shaftless screw design.

DISC SCREENS

A disc screen consists of a flat disc covered with screening media that rotates about a horizontal axis, perpendicular to the flow. Influent enters the submerged portion of the disc, and solids are retained by the screening media. The rotation of the disc lifts the solids above the water surface where they are removed by a bank of spray nozzles.

Disc screens are available in sizes ranging from 6 to 16 ft (1.8 to 5 m) in diameter with mesh openings as fine as .040″ (1 mm).

Another type of disc screen uses one or more pairs of slowly rotating, mesh-covered discs to accomplish removal of fibrous solids. As the discs rotate, solids form a "precoat" on the surface of the discs. The mat of solids concentrates and rolls back by gravity as the discs rotate to self-clean the mesh. The debris exits the unit by means of a debris chute located between each pair of discs.

This screen is available with openings to 44 microns and disc diameters to 60″ (1524 mm). Drive motor sizes range from 1/3 to 5 horsepower (0.25 to 3.7 kW).

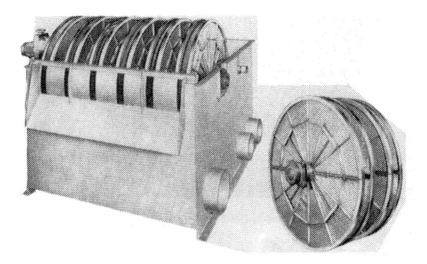

Disc screen, courtesy of Hycor Corp.

VERTICAL DRUM SCREEN

A vertical drum screen is a mechanically cleaned screening device designed for installation in a 25″ to 36″ (635 to 915 mm) comminuter chamber. The vertically oriented screen drum is positioned over a u-tube that discharges to an outlet channel. Incoming flow is directed through the screen and down into the discharge pipe while solids are retained on the face of the screen cylinder.

Cleaning is accomplished by a stationary, toothed cleaning rake, or comb. As solids are removed from the face of the drum, they are directed to a screenings lift that elevates them out of the flow and discharges them into an appropriate container.

This screen is available with screen openings of 1/8″ or 3/16″ (3 or 5 mm), and is operated by a hydraulic power pack typically equipped with a 5 horsepower (3.7-kW) motor.

BRUSH RAKED FINE SCREEN

Brush raked fine screens utilize revolving bristle brushes to remove solids from a curved stationary screen. The cleaning brush mechanism is mounted on a drive shaft that operates in pillow block bearings and slowly rotates in a complete 360-degree circle about a horizontal axis. The rotation of the brush sweeps clean the upstream face of the screen and elevates

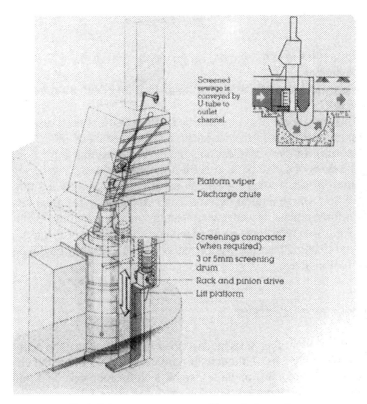

Screened sewage is conveyed by U-tube to outlet channel.

Platform wiper
Discharge chute

Screenings compactor (when required)
3 or 5mm screening drum
Rack and pinion drive
Lift platform

Vertical drum screen, courtesy of Jones & Attwood.

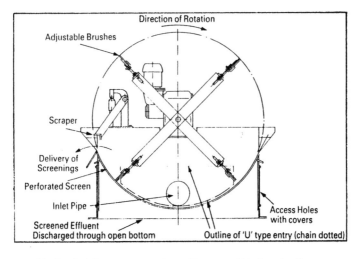

Direction of Rotation

Adjustable Brushes

Scraper

Delivery of Screenings

Perforated Screen

Inlet Pipe

Screened Effluent
Discharged through open bottom

Access Holes
with covers

Outline of 'U' type entry (chain dotted)

Brush raked fine screen, courtesy of Longwood Engineering Corp.

181

solids to the top of the screen, for gravity discharge into a screening trough or container.

Horizontal shaft units may be up to 15 ft long with a screen radius of 3 ft (1 m).

This screen can also be furnished with a vertical shaft design for river intakes up to 16 ft (5 m) deep with a screen radius of 18" to 60" (460 to 1525 mm). In these applications, the curved screening surface is recessed in the sides of a channel that is flush mounted with the river bank. As the rotating brushes clean the screen surface, debris is carried out into the river flow where natural currents can carry it away from the intake.

Perforated plate or wedgewire openings may range from 1/32" to 3/4" (1 to 19 mm). Screens may be fitted with one or two sets of nylon or polypropylene cleaning brushes. Motor/reducer drive units may be direct-coupled or shaft-mounted.

OTHER FINE SCREENS

Variations of other screens described in this book may be adapted to be used as fine screens. Although they are usually used where the clear openings are 1/4" (6 mm) or larger, some reciprocating rake and multi-raked units have been reported to be successful in some fine screening applications. When such a screen is adapted to be used as a fine screen, precautions should be taken to insure that it will be adequately sized to handle the required flow and the increased volume of solids.

CHAPTER 8

Screenings Handling and Disposal

As the use and dependence on screens has increased, so has the concern over the handling and disposal of the screenings they remove. Many landfills can no longer accept screenings because of their high moisture and/or organic content, and the disposal or incineration costs of large volumes of raw screenings may be prohibitive.

The solids, or screenings removed by bar screens and fine screens in wastewater treatment plants are usually putrescible and offensive, and it is necessary that they are handled and disposed of properly to minimize spillage, health, and odor problems.

Even the relatively innocuous screenings from traveling water screens, trash rakes, and other raw water intake screens can no longer simply be returned to the river or lake from where they came.

The solution to a screenings problem requires a well-thought-out and integrated plan for efficient screenings handlings and reuse or disposal. A variety of mechanical devices are available to dewater, convey, compact, and bag screenings.

CONVEYING SYSTEMS

As they are removed from the screen, screenings are usually discharged into a container or trough. Multiple screen installations may be designed to share a common trough, or a conveyor, to consolidate screenings at a central point for further handling and disposal. A screenings trough will require a steep incline, usually greater than forty-five degrees, or a water sluicing system to insure that debris will not accumulate under the discharge point of the screen.

Selection of the most appropriate type of conveyor system will depend on the volume, moisture content, and consistency of the screenings, as well as the distance and elevation of the final destination. The most simple

Bar screen with screenings handling and bagging system, courtesy of EMU GmbH.

In-channel fine screen with screenings compactor, courtesy of Hercules Systems.

Common screenings trough with compactor and conveyor, courtesy of Zenith.

Common screenings belt conveyor, courtesy of Serpentix Conveyor Corp.

and inexpensive conveying device is the belt conveyor. Belt conveyors consist of an endless wire mesh or rubber belt operating over head and tail pulleys. Belts may be troughed or equipped with flights to increase debris-carrying capacity and prevent solids from falling off the belt. Belt conveyors should be furnished with a spray wash system to clean the belt and prevent a buildup of solids.

Wet screenings will undergo some dewatering while being conveyed, as free water gravity drains onto the belt. Therefore, provisions must be made to collect this water and return it to the channel.

Most belt conveyors operate horizonally or at a slight inclination and have limited ability to change directions. However, some proprietary belt conveyors are made with troughed belts and/or convolutions that enable the conveyor to change elevations and directions while conveying wet materials. As the belt passes over the discharge terminal, it is stretched flat so that it can be scraped clean.

Screw conveyors consist of a helical screw operating within a semi-enclosed trough. As the screw is rotated, solids are conveyed along the length of the screw. Most units are equipped with a draining zone to facilitate the separation of liquids and solids. Screw conveyors can be designed to operate horizontally, vertically, and at an inclination. Some designs utilize a shaftless screw to increase conveying capacity and reduce the possibility of debris winding around the shaft.

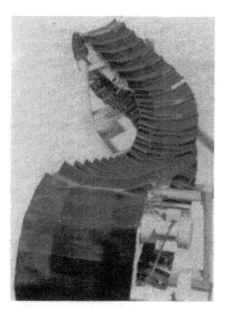

Troughed-belt serpentine conveyor, courtesy of Serpentix Conveyor Corp.

Pneumatic handling system ejector vessel, courtesy of Wheelabrator/CPC.

Conveyor systems can be electrically interlocked to operate when a screen is operated, or at preprogrammed intervals.

Pneumatic conveying equipment can be used to convey screenings from wastewater treatment plants up to 1000 ft (300 m) or more. Screenings are discharged into an ejector vessel where compressed air is introduced to move material through a discharge pipeline 6″ to 8″ (152 to 203 mm) in diameter. The use of enclosed pipes to transport screenings reduces problems associated with spillage and odors and allows horizontal or vertical directional changes. However, pipelines should be designed with long-radius turns and provisions to disassemble for cleaning. A variation on this system utilizes a vacuum pump to convey screens through a pipeline into a vacuum tank.

SCREENINGS WASHERS

Screening solids removed from bar screening equipment at a wastewater treatment plant often contains fecal and other organic matter that readily decomposes. Several devices have been specifically designed to break up

Screenings washer, courtesy of Jones & Attwood.

Screenings washer, courtesy of Steinmann + Ittig.

fecal solids and wash organic matter from the screenings so that it can be returned to the plant flow for further treatment, preventing odor, aesthetic and other problems relating to unsanitary handling conditions. An effective screenings washer will also reduce the weight of screenings by up to 80% and reduce their volume by up to 50%.

Washing is usually accomplished through air or water mixing, liquefaction, maceration, or milling followed by a water rinse.

Screenings washers are most effective when used in conjunction with a screenings compactor.

SCREENINGS COMPACTING

Facilities for disposal of screenings are becoming more expensive and more difficult to locate. It is often essential that the volume of screenings be reduced to a minimum for economic and/or regulatory reasons.

Screw- and piston-type screenings compactors can be used to continuously and effectively dewater, compress, and convey screenings that are free of sticks and other large objects. Depending on the application and screenings consistency, wet screenings can be pressed to 20% to 50% dry solids and a volume reduction of up to 75%. In addition, the compacted screenings may be automatically conveyed up to 30 ft (9 m).

In a screw press, screenings are introduced directly onto the rotating screw and transported to a compaction zone where increased resistance results in their compaction and the formation of a relatively dry plug of solids. A drainage system insures that the liquid removed during conveying and compacting is channeled away from the solids and out of the unit. Many of the screw presses utilize a shaftless screw to reduce the risks of clogging with fibrous and sticky materials.

Operation of the hydraulic piston press is similar to the screw press; however, this press utilizes a hydraulically actuated or motorized piston-type ram to push screenings into the compaction zone.

For information on screens that incorporate screw-type compactors as an integral part of their design, see the description of in-channel rotary screens in Chapter 7.

SCREENINGS GRINDING

Some plants use grinders or macerators to grind and chop screenings into particle sizes of 1/4" to 3/8" (6 to 9 mm). The particles can then be pumped to a fine screen for dewatering and/or compacted in a screenings compactor.

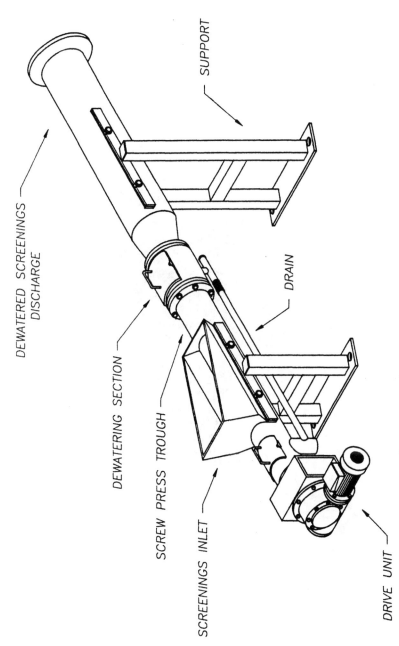

DEWATERED SCREENINGS
DISCHARGE

SUPPORT

DEWATERING SECTION

DRAIN

SCREW PRESS TROUGH

SCREENINGS INLET

DRIVE UNIT

Screw-type screenings compactor, courtesy of Arlat, Inc.

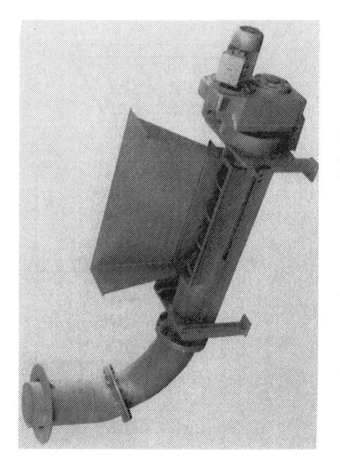

Screw-type screenings compactor, courtesy of Infilco Degremont.

Hydraulical piston-type screenings compactor, courtesy of Roto-Sieve AB.

Screenings bagging unit, courtesy of EMU.

Multi-station bagging unit, courtesy of Longwood Engineering.

194

Multiple station screenings container, courtesy of Geiger GmbH & Co.

Integrated screen/macerator/dewatering systems are available in a single compact unit for small plants with relatively small flows.

See Chapter 9 for additional information on the application and operation of grinders.

BAGGING

Dewatering screenings can be automatically bagged to provide a means of convenient, odorless storage before further handling or incineration. Bagging devices are usually used in conjunction with screenings dewatering/compacting units. Simple, single-station bagging units may consist of a bagging adapter fitted to the discharge end of the compacting device. As bags fill, an operator must seal the bag and pull a new one into place.

Fully automatic units are also available. One such unit consists of multiple bagging stations that monitor the level/weight of the bag filled with an adjustable infrared sensor. When a bag fills to its 110-lb (50-kg) capacity, the unit indexes to an empty bagging station. As the last bag begins to fill, an audible alarm is sounded to alert the plant operator.

CHAPTER 9

Comminuters and Grinders

AS an alternative to screening systems that remove solids from influent steams, some wastewater treatment plants use either comminuters or grinders to reduce solids into particles ranging in size from 1/4" to 3/4" (6 to 19 mm) without removing them from the flow.

Because the solids are not removed from the influent, problems associated with the handling and disposal of screenings are reduced, and organic matter can be biodegraded by secondary treatment processes. The smaller, more uniform particles are less likely to clog pumping equipment and are more suitable for treatment in sedimentation and digestion processes.

However, solids from comminuters and grinders can cause problems in downstream processes if deposits of plastic or other inorganic matter are allowed to accumulate in aeration and digestion tanks. Ground rags and other solids may form "ropes" or "balls" of material that can clog aerators, mixers, pumps, pipes, and other equipment.

Grit and other solids can result in severe wear on grinders, and routine inspection of the cutter teeth and bearings is required.

Because of their reduced open area, comminuter and grinder installations should allow for a headloss greater than that of a bar screen. Based on the flow and clear opening, headlosses typically range from 2" to 12" (50 to 305 mm), although they can reach as high as 3 ft (1 m) or more during periods of maximum flows.

COMMINUTERS

The comminutor is an automatic screening and shredding device that is used to continuously screen large solids and shred them into smaller particles. A comminutor consists of a slotted drum that slowly rotates on a vertical axis. The drum is a combination screen and support for the

197

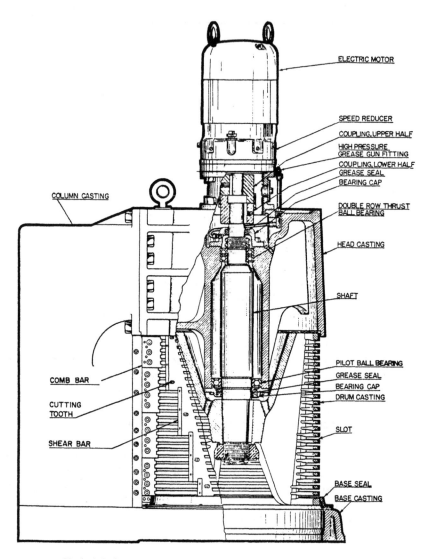

ELECTRIC MOTOR

SPEED REDUCER
COUPLING, UPPER HALF
HIGH PRESSURE
GREASE GUN FITTING
COUPLING, LOWER HALF
GREASE SEAL
BEARING CAP

DOUBLE ROW THRUST
BALL BEARING

HEAD CASTING

SHAFT

PILOT BALL BEARING
GREASE SEAL
BEARING CAP
DRUM CASTING

SLOT

BASE SEAL

BASE CASTING

COLUMN CASTING

COMB BAR

CUTTING
TOOTH

SHEAR BAR

Vertical shaft comminuter arrangement, courtesy of Yeomans Chicago.

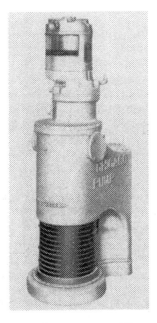

Vertical shaft comminuter, courtesy of Yeomans Chicago.

peripheral cutting teeth and shear bars and ranges in size from 4″ to 36″ (102 to 914 mm) diameter. As the drum rotates, solids too large to pass through the screen slots are cut by the teeth until they are small enough to pass through the slot openings.

Comminutors have been installed in lift stations, small wastewater treatment plants, and some industrial applications, with varying degrees of success. They are usually used to reduce odors, flies, and the unsightliness of screenings removed by other methods or to eliminate screenings and screenings handling completely. Because their cutting teeth are not able to cut hard objects, including stone and metal particles, comminutors are usually not recommended for combined sewer flows.

Comminutors may be located in open channels for straight-through flow or at the end of a channel with a bottom discharge. Some in-channel comminutors are available for mounting on an autocoupling slide rail system to allow quick removal from wet well.

Pipe-mounted comminutors can be hinged to allow access of cutting elements for inspection and maintenance. The cutting chamber is sealed with a neoprene o-ring.

Comminutors are available with slot widths of 1/4″ to 3/8″ (6 to 9 mm) and drum diameters of 7″ to 54″ (178 to 1372 mm). Motor sizes typically

Vertical shaft comminuter arrangement, courtesy of Yeomans Chicago.

range from 1/4 horsepower (0.2 kW) on small units to 5 horsepower (3.7 kW) on larger units.

Submersible drives are offered for operation under flooded conditions, and drive shaft extensions are available for below-grade installations.

Another comminutor design utilizes a conical screen rotor carried out by a central shaft and rotating about its horizontal axis. Hardened steel cutters are mounted around the periphery of the rotating screen to provide a cutting action against stationary knives fixed to the machine frame. This machine can be hydraulically or electrically driven.

A combination bar screen and comminuting device is also available. This unit is equipped with a rotating cutter that travels up and down the face of a stationary bar rack. The cutter rides freely on guide rails and can swing away from the screen if noncomminutable solids are encountered. The cutter is raised and lowered using a wire rope, or telescoping screw arrangement. The direction of rotation of the cutting unit is alternated to prevent debris from accumulating. This unit is available with 3/8″ to 3/4″ (9 to 19 mm) bar spacings for channel measuring 1 to 8 ft (0.3 to 2.4 m) wide.

GRINDERS/MACERATORS

Grinders, or macerators, consist of two sets of shaft-mounted, counter-rotating, and intermeshing cutters that trap solids and chop, rather than shred, them into smaller particles. The vertical shafts rotate at different speeds to produce a self-cleaning action of the cutters.

Grinders aggressively chop solids into particle sizes of 1/4″ to 3/8″ (6 to 9 mm), and they can be configured for use in open channels up to 5 ft (1.5 m) deep or pipelines up to 12″ (305 mm) diameter. Units can be operated by means of electric or hydraulic drive units.

A modular grinder/screen unit is available for applications with high flows or wide influent channels. This unit utilizes a horizontally rotating

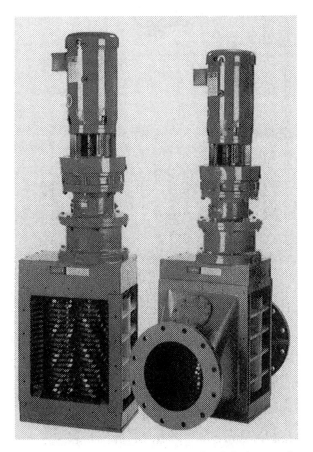

In-channel and in-line grinders, courtesy of JWC Environmental.

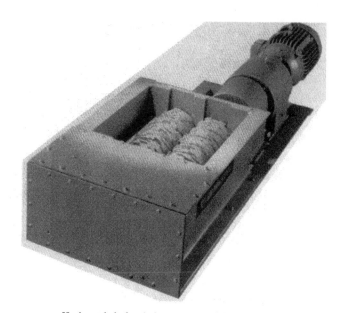

Horizontal shaft grinder, courtesy of Franklin Miller.

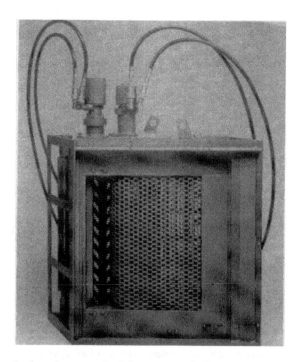

In-channel screen and grinder, courtesy of JWC Environmental.

202

screen that diverts solids to a grinder located at one side of the channel. This grinder/screen unit can be designed to discharge the ground particulates back into the flow, or a solids removal system can be furnished to collect the ground solids on the downstream side of the grinder and pump them to a dewatering device for disposal.

CHAPTER 10

Microscreens

MICROSCREENS, or microstrainers, are mechanical filtration devices that employ a fine mesh fabric mounted on the periphery of a horizontally mounted, continuously rotating drum. Approximately 70 to 75% of the drum (66% of the screening area) is submerged under normal operation. Flow enters the drum interior through the open end and exits radially through the screening mesh while solids are retained on the inner surface of the mesh. Captured solids are washed from the screening mesh into a backwash trough using a series of spray nozzles located at the top of the screen.

Microscreens are equipped with filter mesh having openings that range from 1 to 200 microns, although most applications utilize openings within a 15- to 60-micron range. Screening is accomplished directly by capturing a solid on the screening media or indirectly by capturing a solid on a "mat" or thin film of solids previously captured on the mesh surface.

The indirect capture of solids contributes to the efficiency of the screening process by reducing the effective size of the screening mesh. Because microscreens are equipped with such fine mesh openings, careful mechanical and process considerations must be utilized in their application, design, and operation.

APPLICATIONS

Microscreens have been used in a variety of water and wastewater applications for more than thirty years. Increasing technology is renewing interest in the application of microcreens to meet stricter operating conditions.

Water treatment plants using a surface water supply may use microscreens as an initial treatment step to remove a significant proportion of raw water suspended solids. This may include the removal of up to 75%

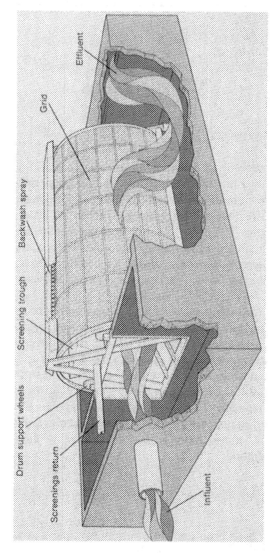

Effluent

Grid

Backwash spray

Screening trough

Drum support wheels

Screenings return

Influent

Microscreen flow diagram, courtesy of Envirex, Inc.

206

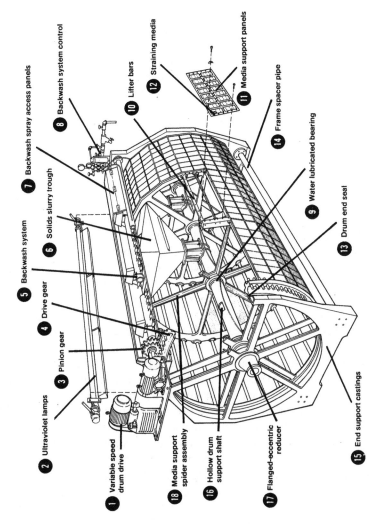

1 Variable speed drum drive

2 Ultraviolet lamps

3 Pinion gear

4 Drive gear

5 Backwash system

6 Solids slurry trough

7 Backwash system

8 Backwash spray access panels

9 Water lubricated bearing

10 Lifter bars

11 Media support panels

12 Straining media

13 Drum end seal

14 Frame spacer pipe

15 End support castings

16 Hollow drum support shaft

17 Flanged-eccentric reducer

18 Media support spider assembly

Backwash system control

Microscreen assembly, courtesy of Lakeside Equipment.

of the algae and some free-swimming organisms, reducing chemical costs and increasing the performance of downstream treatment processes.

Some industrial process water applications or direct filtration plants may use a microscreen as the only other treatment step in their water treatment facilities.

In wastewater treatment plants, microscreens may be applied as a tertiary or final "polishing" process or in lieu of replacing secondary sedimentation. Effluent from aerated lagoons and oxidations ponds may also be polished with a microscreen prior to discharge. When used to polish secondary effluents, microscreens have been reported to produce effluent qualities as low as 10 mg/L suspended solids and 10 mg/L BOD in some wastewater applications. Microscreens can also be used to recover or reclaim materials from process waste streams in pulp and paper, meat packing, tanning, and other industrial applications.

MESH

The screen mesh is also referred to as media, fabric, or cloth and is the heart of the microscreen. The size of the mesh openings will determine the overall equipment size and system's process performance.

The available size of mesh openings ranges from 1 to 60 microns, although individual manufacturers may have limitations on the minimum mesh openings that they offer. The screening media is most frequently woven of polyester filaments although stainless steel wire is offered by some manufacturers when clear openings are greater than 15 microns.

Mesh durability is important to the success of a microscreen installation. The combination of small openings and fine wire diameters results in a thin, flexible piece of mesh that requires intermediate support. Adequate support is usually provided by installing the mesh in small panels or frames, each having a screening area that ranges from 2-1/2 to 10 ft² (0.25 to 0.93 m²). These flat or involute-shaped panels are then bolted to the microscreen drum with easily removable stainless steel fasteners.

Some microscreen manufacturers offer proprietary panel designs that have been developed to overcome operational problems associated with fine mesh and facilitate panel installation, removal, and replacement. Most of these designs incorporate the use of a molded plastic mesh panel that is manufactured with an integral support grid. The square of hexagonal grid will support the mesh at 1″ to 3″ (25 to 75 mm) centers.

DRUM AND SUPPORT FRAME

Microscreens are available in diameters that range from 6 to 12 ft (2 to 3.7 m) and lengths ranging from 4 to 18 ft (1.2 to 5.5 m). Manufacturers

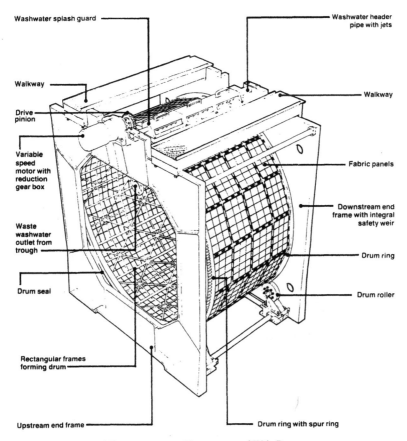

Washwater splash guard

Washwater header pipe with jets

Walkway

Walkway

Drive pinion

Variable speed motor with reduction gear box

Fabric panels

Waste washwater outlet from trough

Downstream end frame with integral safety weir

Drum ring

Drum seal

Drum roller

Rectangular frames forming drum

Upstream end frame

Drum ring with spur ring

Microscreen assembly, courtesy of Weir Pumps.

also offer smaller, "packaged" microscreens in steel tanks for low flow applications.

Design and construction of the drum assembly and support frame varies among equipment manufacturers. The basic mesh support structure is a cylindrically shaped, suitably stiffened rigid frame fabricated of carbon or stainless steel. End supports/plates may be fabricated or cast of steel, stainless steel, or aluminum.

There are several methods used to support the drum assembly. One method utilizes a pair of nylon rollers to support the upper inside surface of the inlet side of the drum. The closed end of the drum is supported by a stub shaft operating in water lubricated thrust bearing. Another manufacturer assembles the end of the drum structure to cast bronze end rings that are supported by tire-like support rollers. Other designs include the

Microscreen frame assembly, courtesy of Hubert Stavoren.

use of a central shaft to support the drum frame and a support arrangement using a large diameter journal bearing located in the end frame to support a bearing flange formed as part of the drum end plate.

Unscreened water is prevented from passing around both ends of the drum by a flexible seal. At least one manufacturer offers a seal arrangement that can compensate for wear and can automatically increase sealing pressure in response to an increase in headloss.

BACKWASH SPRAY SYSTEM

Debris is removed from the drum by means of a pressurized water spray. A spray pipe header equipped with replaceable nozzles is located above the drum, directing a water spray through the mesh and into a backwash trough located inside the drum. The backwash water is usually obtained from the screened microscreen effluent. Backwash water requirements may range from 2 to 5% of the screen's total throughput, at a pressure of 15 to 70 psi (1 to 4.8 bar).

Spray systems may be designed and operated to suit a particular, or varying, operating condition. Options include single or dual spray pipes, for constant or intermittent operation, at a constant or variable pressure.

Periodic spray nozzle purging may be required. It is recommended that self-cleaning spray nozzles be utilized to reduce maintenance and increase process reliability. The spray system is covered with a corrosion-resistant spray hood to prevent overspray. Most manufacturers offer hoods equipped with clear plastic inspection windows for visual observation of the operating spray system. A walkway or bridge is recommended to provide access for maintenance and inspection of the spray system.

The backwash trough, located at the inside top of the drum, should be designed with steeply sloping sides to prevent debris accumulations.

SUPPLEMENTAL MESH CLEANING

Over a period of time the microscreen mesh may become clogged with oil, grease, or biological growths. Supplemental mesh cleaning will be required to reduce this clogging. Depending on the application, manufacturers suggest cleaning the mesh at intervals that range from every three to eight weeks of operation.

Several cleaning methods are available based on the type of fouling that has occurred. When selecting a supplemental cleaning system, its compatibility with the mesh material should be considered. These cleaning methods include the use of mild chlorine solutions, ultraviolet irradiation, steam, and high-pressure hot or cold water sprays.

Each cleaning system has advantages and disadvantages. The use of chlorine will require that the screens be removed from service for up to eight hours or more. Chlorine also has a deleterious effect on stainless steel media. Ultraviolet irradiation may require frequent lamp replacement and is effective only on the exposed surfaces of the biological growths. UV irradiation is also harmful to polyester mesh. High-pressure water or steam cleaning is also expensive and time consuming.

DRIVE AND CONTROLS

Microscreens are rotated at a peripheral drum speed that ranges between 15 to 150 fpm (4.6 to 46 m/min). Rack and pinion, drive chain and sprocket, and v-belt and sheave arrangements may be used to transfer power from an electric motor-speed reducer drive unit to the drum.

A rack and pinion drive gear system consists of a nylon or bronze pinion gear keyed to the output shaft of the speed reducer. The pinion engages a larger gear ring fixed to the periphery of the drum's drive end. Designs utilizing segmented cut tooth gears and pin rack type gear rings are available.

Power can also be transferred to the drum by means of a drive chain/v-belt operating over a drive and driven sprocket/sheave. The drive sprocket is mounted on the output shaft of the reducer, and the driven sprocket is mounted to a stub or central shaft fixed to the drum.

The drum's rotational speed should be adjustable and may be automated to accommodate sudden variations in raw water quality and quantity. Variable speed may be accomplished using a DC drive motor, variable frequency AC drive motor, or transmission.

In most applications it is preferred that a constant headloss across the screen be maintained. To accomplish this, the screen control system should be capable of varying the rotational speed of the drum as the differential headloss across the screen increases or decreases. Rotational speed can be controlled using the output signal from a pneumatic- or capacitance-type differential level controller. Additional controls can be furnished to regulate backwash operation and provide other specific process requirements.

SIZING AND PROCESS CONSIDERATIONS

The mechanical process of microscreening is fairly simple. However, it is very hard to predict the effectiveness of the process for a specific application. Microscreens are usually sized on the basis of hydraulic loading of the submerged screen area (i.e., gpm/ft^2) at a given headloss. Determination of the loading for a specific application is made based on a thorough review of the application, data from similar installations, and/or pilot- and bench-scale testing.

Typical hydraulic loadings range from 5 to 20 gpm/ft² of submerged area at 12″ of headloss. However, units with openings of 1 to 7 microns may be rated at 1.5 gpm/ft² at a headloss of up to 24″. Industrial applications and those with larger mesh sizes may utilize loadings of 50 gpm/ft² or more.

Most manufacturers have developed filterability test kits that can be used to determine the effectiveness of a microscreen for individual applications. These field test kits can be used as guides in the selection of mesh size and hydraulic loading and can reasonably predict effluent quality.

All pilot testing should be done during periods of maximum solids loading and simulate actual operation (i.e., location, temperature, pretreatment, etc.) as closely as possible.

Screening Equipment Manufacturers

THE following list of manufacturers includes the company name, address, phone and fax numbers, and the type of screens that they currently offer. It is recommended that the manufacturer be contacted to confirm the range of their products.

Many of the listed manufacturers have subsidiaries, affiliates, and distributors or agents in countries not listed below.

B = Bar Screens
C = Comminutors, Grinders
D = Drum/Disc Screens
F = Fine Screens
M = Microscreens
P = Passive Screens
R = Trash Rakes
S = Screenings Dewatering/Disposal
T = Traveling Water Screens

Advanced Wastewater Treatment Ltd.   F
Lynwood House, Dudley Road, Lye
Stourbridge, West Midlands DY9 8DU
England
Phone: 0384-892619
Fax: 0384-424081

Aer-O-Flo Environmental, Inc.   F
1175 Appleby Line, Unit B-2
Burlington, Ontario
Canada L7L 5H9
Phone: 416-335-8944
Fax: 416-335-8972

ALPHA Umwelttechnik AG   **B, F, T**
Schloss-Strasse 15
CH-2560 Nidau
Switzerland
Phone: 032-51-54-54
Fax: 032-51-23-27

Amwell, Inc.   **B**
1740 Molitor Road
Aurora, Illinois 60505-9990
Phone: 708-898-6900
Fax: 708-898-1647

Andritz Sprout-Bauer, Inc.   **F**
Sherman Street
Muncy, Pennsylvania 17756
Phone: 717-546-8211
Fax: 717-546-1306

Andritz Sprout-Bauer, S.A.   **F**
10, Avenue de Concyr
45071 Orleans Cedex 2
France
Phone: 38-51-5738
Fax: 38-63-1565

Aquacare Environmental, Inc.   **M**
1155 N. State Street #303
Bellingham, Washington 98225
Phone: 206-734-7964
Fax: 206-734-9407

Arlat, Inc.   **B, F, S**
150 East Bramalea
Ontario
Canada L6T 1C1
Phone: 416-458-8220
Fax: 416-458-8224

Atlas Polar Co., Ltd.   **R**
Hercules Division
P.O. Box 160, Postal Station "O"
Toronto
Canada M4A 2N3
Phone: 416-751-7740
Fax: 416-751-6475

Awatech GmbH   F
Postfach 120 263
30908 Isernhagen
Germany
Phone: 0511-72898-0
Fax: 0511-7298-71

Bamag GmbH   **B, D, R, T**
Postfach 460
Wetzlarer Straße 136
D-6308 Butzbach
Germany
Phone: 060-33-839
Fax: 060-33-83-506

E. Beaudrey & Company   **B, D, R, T**
14 Boulevard Ornano
75018 Paris
France
Phone: 42-57-14-35
Fax: 42-64-74-62

Bieri Hydraulik   **R**
Könizstrasse 274
CH-3097 Liebefeld
Switzerland
Phone: 031-971-0973
Fax: 031-972-1049

Biwater Treatment Ltd.   **B, S**
Biwater Place, Gregge Street
Heywood, Lancashire
OL10 2DX
England
Phone: 0706-367555
Fax: 0706-365598

Bormet Maschinenbau GmbH   F
Postfach 148
6108 (64331) Weiterstadt
Germany
Phone: 061-50-5098-0
Fax: 061-50-5098-31

Brackett Green Ltd.   **B, D, F, M, R, S, T**
Severalls Lane, Colchester
Essex CO44PD
England
Phone: 0206-852121
Fax: 0206-844509

Brackett Green USA, Inc.   **B, D, F, M, R, S, T**
16850 Saturn Lane, Suite 140
Houston, Texas 77058
Phone: 713-480-7955
Fax: 713-480-8225

C&H Waste Processing   **C**
9 Garrett Road
Lynx Trading Estate, Yeovil
Somerset, BA20 2TJ
England
Phone: 0935-26927

Chicago Pump Products,   **C**
Yeomans Chicago Corp.
1999 North Ruby Street
Melrose Park, Illinois 60160
Phone: 708-344-4960
Fax: 708-681-4432

Claude Laval Corp.   **P**
Lakos Filtration Systems
1911 N. Helm
Fresno, California 93727
Phone: 209-255-1601
Fax: 209-255-8093

Contec GmbH   **F**
Postfach 6148
53594 Bad Honnef
Germany
Phone: 02224-80496
Fax: 02224-81426

Contra-Shear Engineering Ltd.   **D, F, S**
CPO Box 1611
Auckland
New Zealand
Phone: 09-818-6108
Fax: 09-818-6599

Cook Screen Technologies, Inc.   **P**
1292 Glendale-Milford Rd.
Cincinnati, Ohio 45125
Phone: 513-771-9192
Fax: 513-771-2665

Copa Group   **F**
Crest Industrial House Estate, Marden
Tonbridge, Kent, TN12 9QJ

England
Phone: 44-0622-832444
Fax: 44-0622-831466

COS.M.E.   **B, F, S**
Via Torino, 6 (fraz. Marola)
36040 Torri Di Quartesolo (Vicenza)
Italy
Phone: 0444-582566
Fax: 0444-582185

Cross Machine, Inc   **R**
167 Glen Avenue
Berlin, New Hampshire 03570
Phone: 603-752-6111
Fax: 603-752-3825

Diemme USA   **S**
1866 Colonial Village Lane, Ste 111
Lancaster, Pennsylvania 17601
Phone: 717-394-1977
Fax: 717-394-1973

Dontech, Inc.   **B, F, S** ´
76 Center Drive
Gilberts, Illinois 60136
Phone: 708-428-8222
Fax: 708-428-6855

Dorr-Oliver, Inc.   **B, F, M**
P.O. Box 3819
Milford, Connecticut 06460-8719
Phone: 203-876-5400
Fax: 203-876-5432

Dorr-Oliver Co., Ltd.   **B, F, M**
NLA Tower
12/16 Addiscombe Road
Croydon, CR9 2DS
England
Phone: 081-686-2488
Fax: 081-681-8655

Dresser Pump   **C**
Waste-Tec Operations
Horsfield Way, Bredbury Park
Stockport SK6 2SU
England
Phone: 061-406-7111
Fax: 061-406-7222

Duperon Corporation   F
  5693 Becker Road
  Saginaw, Michigan 48601
  Phone: 517-754-8800
  Fax: 517-754-2175

E&I Corporation   B, T
  6805 Oak Creek Drive
  Columbus, Ohio 43229
  Phone: 614-899-2282
  Fax: 614-899-0304

E.M.O. France (Siège Social)
  B.P. 22
  35740 Pacé
  France
  Phone: 99-60-6363
  Fax: 99-60-2087

EMU Umwelttechink   F, S
  August-Mohl Straße 38
  D-95030 Hof (Saale)
  Germany
  Phone: 09281-974520
  Fax: 09281-974593

Enviro-Care Company   B, C, S
  5614 W. Grand Avenue
  Chicago, Illinois 60639
  Phone: 312-745-7773
  Fax: 312-745-8383

Environmental Engineering Ltd.   F
  Little London, Spalding
  Lincolnshire RE11 2UE
  England
  Phone: 0775-768964
  Fax: 0775-710294

Enviroquip, Inc.   B
  P.O. Box 9069
  Austin, Texas 78766
  Phone: 512-218-3200
  Fax: 512-218-3277

Envirex, Inc.   B, D, M, R, T
  1901 South Prairie Avenue
  Waukesha, Wisconsin 53186
  Phone: 414-547-0141
  Fax: 414-547-4120

Epco International Pty. Ltd.   **C**
29A Capella Crescent
Moorabbin, 3189
Victoria
Australia
Phone: 03-555-2688
Fax: 03-555-8052

Fairfield Service Company   **B, R, T**
240 North Boone Avenue
Marion, Ohio 43302
Phone: 614-387-3335
Fax: 614-387-4869

FMC Corporation   **B, C, D, F, R, S, T**
Material Handling Systems Division
400 Highpoint
Chalfont, Pennsylvania 18914
Phone: 215-822-4300
Fax: 215-822-4342

F.P.I., Inc.   **T**
P.O. Box 1477
Shafter, California 93263
Phone: 805-589-6901

Franklin Miller, Inc.   **B, C**
60 Okner Parkway
Livingston, New Jersey 07039
Phone: 201-535-9200
Fax: 201-535-6269

FRC Environmental, Inc.   **F**
P.O. Box 2453
Gainesville, Georgia 30503
Phone: 404-534-3681
Fax: 404-535-1887

Friedrich Schrage   **B**
Badenstedter Str 98 A + B
3000 Hannover 91
Germany
Phone: 0511-494091
Fax: 0511-498892

FSM – Frankenburger   **B, F**
6301 Pohleim 2
Garbenteich
Germany
Phone: 06404-2265/61533
Fax: 06404-63437

Geiger GmbH & Co.   **B, D, R, S, T**
P.O. Box 210163
76151 Karlsruhe 21
Germany
Phone: 0721-5001-285
Fax: 0721-5001-370

G.E.T. Industries, Inc.   **C**
P.O. Box 640, Brampton
Ontario L6V 2L6
Canada
Phone: 416-451-9900
Fax: 416-451-5376

Hans Huber GmbH   **B, F**
Huber Edestahl
Mariahilfstraße 3–5
D-8434 Berching
Germany
Phone: 08462-201-0
Fax: 08462-201-4

Hans Künz GesmbH   **R**
A-6971 Hard
Austria
Phone: 43-5574-38683-0
Fax: 43-5574-38683-19

Hendrick Screen Co.   **F, P**
P.O. Box 369
Owensboro, Kentucky 42302-0369
Phone: 502-685-5138
Fax: 502-685-1729

Hoelschertechinc-Gorator GmbH   **F**
Postfach 1426
Am Trimbuschof 20
4690 Herne
Germany
Phone: 02323-510-6769
Fax: 02323-18098

Hubert Stavoren BV   **B, D, M, R, T**
Burg, Stramanweg 108
Postbus 12520
1100 AM Amsterdam Z-O
Phone: 020-977-699
Fax: 020-979-959

Hycor Corporation   **B, D, F, S**
29850 North Highway 41
Lake Bluff, Illinois 60044-1192
Phone: 708-473-3700
Fax: 708-473-0477

Hydro-Aerobics, Inc.   **B, F**
1615 State Route 133
Milford, Ohio 45150
Phone: 513-575-2800
Fax: 513-575-2896

Hydro Group, Inc.   **P**
Ranney Division
2 North State Street
Westerville, Ohio 43081
Phone: 614-882-3104
Fax: 614-882-3071

Hydropower Turbine Systems   **B, R**
2805 Woodmark Court
Richmond, Virginia 23233-1693
Phone: 804-360-7992
Fax: 804-360-7993

Hydropress Wallander & Co. AB   **B, F, S**
Heljesvägen 4, Box 125
S-437 22 Lindome
Sweden
Phone: 4631-995050
Fax: 4631-995133

Hydrotech AB   **F**
Vaster 38, Almhaga
S-23591 Vellinge
Sweden
Phone: 040-426020

Hydrothane Systems   **R**
417 U.S. Route 1
Falmouth, Maine 04105
Phone: 207-781-4431
Fax: 207-781-2161

Infilco Degremont, Inc.   **B, R, S**
P.O. Box 29599
Richmond, Virginia 23229
Phone: 804-756-7600
Fax: 804-756-7643

IPEC Industries   F
P.O. Box 1607
Vashon, Washington 98070
Phone: 206-292-8198
Fax: 604-929-7839

Jones and Attwood, Inc.   B, D, F, M, S
1931 Industrial Drive
Libertyville, Illinois 60048-9738
Phone: 708-367-5480
Fax: 708-367-8983

Jones and Attwood, Ltd.   B, D, F, S
Titan Works, Stourbridge
West Midlands, DY8 4LR
England
Phone: 0384-371937
Fax: 0384-392181

JWC Environmental   C
16802 Aston Street, Suite 200
Irvine, California 92714
Phone: 714-833-3888
Fax: 714-833-8858

Kason Corporation   F
1301 E. Linden Ave.
Linden, New Jersey 07036
Phone: 908-486-8140
Fax: 908-486-8598

Köpcke Industrie B.V.   F, S
REKO Products
Delta Industrieweg 36
3251 LX, Stellendam
The Netherlands
Phone: 01879-2988
Fax: 01879-2781

KRC (Hewitt) Inc.   F, M
P.O. Box 68
Neenah, Wisconsin 54957
Phone: 414-722-7713
Fax: 414-725-8615

I. Krüger, Inc.   B
401 Harrison Oaks Blvd.
Cary, North Carolina 27513
Phone: 919-677-8310
Fax: 919-677-0082

Krüger   **B**
Gladsaxevej 363
DK-2860 Soborg
Denmark
Phone: 45-39-690222
Fax: 45-39-690806

Lakeside Equipment Corporation   **B, F, M**
P.O. Box 8448
Bartlett, Illinois 60103
Phone: 708-837-5640
Fax: 708-837-5647

Landustrie Sneek BV   **T**
P.O. Box 199
86 AD Sneek
The Netherlands
Phone: 31-05150-11411
Fax: 31-05150-12398

Lockertex   **F**
P.O. Box 161
Warrington, Chesire WA1 2SU
England
Phone: 0925-51212
Fax: 0925-444386

Longwood Engineering Co., Ltd.   **B, R**
Parkwood Mills, Longwood
Huddersfield, West Yorkshire
HD3 4RP
England
Phone: 0484-642011
Fax: 0484-642935

Lyco, Inc.   **M**
P.O. Box 181
Marlboro, New Jersey 07746
Phone: 908-431-4440
Fax: 908-431-4539

Lyco Manufacturing, Inc.   **F**
P.O. Box 31
Columbus, Ohio 53925
Phone: 414-623-4152
Fax: 414-623-3780

MAHR Maschinebau Ges.mbH   B, S
  Salzgries 1
  1010 Vienna
  Austria
  Phone: 533 55 3127
  Fax: 533 55 3125

John Meunier, Inc.   B, F, R
  6290 Perinault
  Montreal, Quebec
  Canada H4K 1K5
  Phone: 514-334-7230
  Fax: 514-334-5070

Meyn Water Treatment B.V.   F
  Sluispolderweg 44c
  1505 HK Zaandam
  Holland
  Phone: 31-75-313001
  Fax: 31-75-704669

Neuhold Gesellschaft m.b.H   B, R
  Weizer Strasse 49, Postfach 39
  A-8200 Gleisdorf
  Austria
  Phone: 031 12 4057-0
  Fax: 031 12 4320

Nijhuis Water B.V.   B, F
  P.O. Box 43
  7100 AA Winterswijk
  Holland
  Phone: 31-5430-22084
  Fax: 31-5430-20155

Noggerath & Co.   B, F, S
  Feldstrabe 2
  D-31708 Ahnsen
  Germany
  Phone: 057-22-882-0
  Fax: 057-22-882-82

Norair Engineering Corp.   T
  337 Brightseat Road, Suite 200
  Landover, Maryland 20785
  Phone: 301-499-2202
  Fax: 301-499-1342

Optima House Group   F, S
  Hercules Systems
  Askern Road, Toll Bar

Doncaster, South Yorkshire
DN5 0QY
England
Phone: 0302-874128
Fax: 0302-875415

Ossberger Turbinenfabrik GmbH   B, R
Postfach 425
D-8832 Weissenburg/Bavaria
Germany
Phone: 09141-977-0
Fax: 09141-977-20

OVRC Environmental, Inc.   F
Point Corporate Center Drive
Birmingham, Alabama 35243-3331
Phone: 205-969-3010
Fax: 205-969-3020

Parkson Corporation   B, F, S
P.O. Box 408399
Fort Lauderdale, Florida 33340-8399
Phone: 305-974-6610
Fax: 305-974-6182

Passavant-Werke AG & Company   B, R, T
D-6209 Aarbergen 7
Germany
Phone: 49-6120-282434
Fax: 49-6120-282576

Perry Engineering   B
Railway Terrace
Mile End South
Adelaide
South Australia
Phone: 08-3521777
Fax: 08-2341896

Pro-Ent, Inc.   T
P.O. Box 23611
Jacksonville, Florida 32241
Phone: 904-737-3536
Fax: 904-737-3537

Purac Engineering, Inc.   S
5301 Limestone Road, Suite 126
Wilmington, Delaware 19808
Phone: 302-239-9431
Fax: 302-239-9085

Purator Waagner-Brio   B, F
Postfach 53
A-1234 Wien, Lembockgasse 49
Austria
Phone: 1/816 07-0
Fax: 1/816 07-232

Romag AG   B
3186 Dudingen
Switzerland
Phone: 037-436500
Fax: 037-431314

Roto-Sieve AB   F
Box 172
S-442 22 Kungälv
Sweden
Phone: 46(0)303-64210
Fax: 46(0)303-63650

Schloss Engineered Equipment   B, S
Dartmouth Court, Suite 230
10555 E. Dartmouth Ave.
Aurora, Colorado 80014
Phone: 303-695-4500
Fax: 303-695-4507

Schlueter Company   F
P.O. Box 547
Janesville, Wisconsin 53547
Phone: 608-755-0740
Fax: 608-755-0332

Schreiber Corporation, Inc.   B
100 Schreiber Drive
Trussville, Alabama 35173
Phone: 205-655-7466
Fax: 205-655-7669

Schreiber-Klaranlagen GmbH   B
Postfach 1580
30836 Langenhagen
Germany
Phone: 0511-77990
Fax: 0511-7799-22-0

Screening Systems International   B, R, T
P.O. Box 968
Monticello, Mississippi 39654
Phone: 601-587-0522
Fax: 601-587-0524

Sepra Tech, S.A.R.L. F
16, Rue de Vigny
F-68000 Colmar
France
Phone: 8923-8416
Fax: 8923-8432

Sernagiotto F
via Torino 144
27045 Casteggio (PV)
Italy
Phone: (39) 383 83741
Fax: (39) 383 83782

Smalley Excavators, Inc. R
71 Hartford Turnpike South
Wallingford, Connecticut 06492
Phone: 203-265-9352

Spirac AB F, S
Box 30033
200 61 Malmo
Sweden
Phone: 46-040-162020
Fax: 46-040-153650

Spirac/JDV Equipment Corp. F, S
P.O. Box 471
Montville, New Jersey 07045
Phone: 201-335-4740
Fax: 201-335-4702

Steinmann + Ittig F, S
P.O. Box 3110
32425 Minden
Germany
Phone: 0571-4045-0
Fax: 0571-4145-299

Stengelin GmbH & Co. M
Donaueschinger Strabe 52–56
D-7200 Tuttlingen
Germany
Phone: 07461-1795-0
Fax: 07461-179513

SWECO, Inc. F
8029 US Highway 25
Florence, Kentucky 41042
Phone: 606-727-5191
Fax: 606-727-5122

Three Star Environmental Engr.   **B, D, F, R, T**
Platts Common Industrial Estate
Hoyland, Barnsley S74 9TD
England
Phone: 0226-748021
Fax: 0226-748403

Vulcan Industries, Inc.   **B, C, F**
212 South Kirlin Street
Missouri Valley, Iowa 51555
Phone: 712-642-2755
Fax: 712-642-4256

Vollmar GmbH   **B**
Postfach 501148
7000 Stuttgart 50
Germany
Phone: 0711-55391-0
Fax: 0711-5539130

Weir Pumps, Ltd.   **M**
149 Newlands Road, Cathcart
Glasgow G44 4EX
Scotland
Phone: 041-637-7141
Fax: 041-637-7358

Wheelabrator Engineered Systems   **F, P**
Johnson Screens
P.O. Box 64118
St. Paul, Minnesota 55164
Phone: 612-636-3900
Fax: 612-838-3266

Wheelabrator Engineered Systems   **B, C, F, P, S**
Wiesemann Screen Products
P.O. Box 10037
Largo, Florida 34643
Phone: 813-535-4495

Whitehead & Poole Limited   **B**
P.O. Box 9
Radcliffe, Manchester
M26 9NU
England
Phone: 061-723-3821

Zenith Maschinenfabrik GmbH   **R**
Postfach 1160
D5908 Neukirchen/Siegerland
Germany
Phone: 02735-74-0
Fax: 02735-74-211

Zickert Products AB   **B**
Box 53
S-43033 Fjärås
Sweden
Phone: 46(0) 300-33200
Fax: 46(0) 300-33219

Right or Left Hand?

RIGHT-HAND and left-hand terminology is commonly used when discussing screening equipment. These terms are used to describe an entire screen unit, as well as individual screen components. It is important that a universally understood point of reference be used when describing a screen.

Screen Unit: The "right-hand" side of the screen is that side of the screen that is on the right when the screen is viewed from the front, with the viewer's back to the flow.

Screen Chain: Not all chains or attachments are reversible in the direction of travel, in operation over sprockets, or in tracks; therefore, many attachments must be made in both right- and left-hand configurations so they may be applied to conveyors requiring two parallel strands of chain.

Right-hand or left-hand identification is made on the basis of an individual link by holding the attachment link so that the surface that contacts the sprocket or track is down and the closed end of an offset-sidebar chain (or the trailing end of a straight-sidebar link) is away from your body. Under these conditions, if the attachment extends to the right, it is a right-hand attachment; if to the left, it is a left-hand attachment.

Open Area Equations

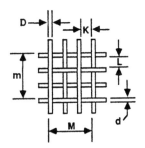

Where:

POA = percent open area
d = shute wire diameter
D = warp wire diameter
K = width of opening (inches)
L = length of opening (inches)
m = shute wire per lineal inch
M = warp wires per lineal inch

For plain, twilled or rectangular:

$$POA = ((K \times L)/(K + D)(L + d)) \times 100$$

or

$$POA = (1 - MD)(1 - md) \times 100$$

To convert meshes per inch to space openings:

$$K = (1/M) - D$$

or

$$L = (1/m) - d$$

To convert space openings to meshes per inch:

$$M = 1/(K + D)$$

or

$$m = 1/(L + d)$$

Selection of Materials

THE selection of the proper materials of construction can be one of the most important aspects of screening equipment design. Many screen failures are attributable to poor material selections of various screen components. With an increasing variety of materials being offered, including over sixty types of stainless steel, the selection of suitable materials of construction can be a challenging assignment.

The selection of a suitable material should involve a review of five important design criteria. These criteria include the fabrication requirements and the material's total cost, corrosion resistance, mechanical properties, and product availability.

This section briefly describes the more common materials and alloys used in the manufacture of water and wastewater screening equipment.

STAINLESS STEELS

This is a general term that is usually used to describe steels that usually contain a minimum of 12% chromium as the principle alloying element. Greater corrosion resistance can be provided by increasing the chromium content up to 27%.

TYPE 302 AND 304

Familiarly know as "18-8 stainless steels," the Type 302 and 304 alloys are popular austenitic stainless steels containing 18% chrome and 8% nickel. They are nonmagnetic steels that cannot be hardened by heat treatment and, instead, must be cold worked to obtain higher tensile strengths. They have excellent corrosion resistance and can be readily fabricated by all methods usually employed with carbon steels. Type 304 contains a maximum 0.08% carbon to make it slightly more corrosion-resistant than Type 302.

TYPE 304L

This is similar to the other 18-8 alloys, except for a lower carbon content of

0.03% maximum, reducing the possibility of carbide precipitation. The corrosion resistance is therefore unaffected by normal welding and stress-relieving applications. Type 304L is usually used if extensive welding is required in fabrication.

TYPE 316

This is a modified version of Type 304, containing 2% to 3% molybdenum. The addition of molybdenum increases the corrosion resistance of this alloy and makes it less susceptible to pitting and pinhole corrosion by seawater, brine, and chlorides.

TYPE 316L

This is similar to Type 316, except for a lower carbon content of 0.03% maximum, reducing the possibility of carbide precipitation. Type 316L is the preferred grade of Type 316 stainless steel if extensive welding is required in fabrication.

TYPE 410

This is a basic 400 series stainless steel alloy containing 12% chrome, which is magnetic and can be readily heat treated to provide a wide range of mechanical properties. It is fairly resistant to mild forms of corrosion.

TYPE 416

This alloy is closely related to Type 410, with similar mechanical properties and slightly lower corrosion resistance. One or more elements usually phosphorous and sulfur, are added to improve machinability.

TYPE 420

This is a 12% chromium steel, with a nominal 0.30% carbon content that results in better responsiveness to heat treatment than Type 410. It must be hardened and ground to obtain maximum corrosion resistance.

TYPE 440

Type 440 is a hardenable 17% chromium steel that offers better resistance to wear than Type 420 and is available in several grades with varying carbon content to provide hardness levels to Rc 60/62.

TYPE 17-4 PH

This martensitic alloy contains 17% chromium and 4% nickel and is usually superior in corrosion resistance to the 400 series stainless steels. 0.04% copper is added to promote precipitation hardening (PH) capabilities.

ALLOY 20

This grade of stainless steel is used for its improved corrosion resistance, especially corrosion by sulfuric acid. It contains approximately 20% chromium and 33% nickel, in addition to small amounts of columbium.

OTHER MATERIALS USED IN SCREEN MANUFACTURE

ABRASION-RESISTANT (AR) STEEL

This is a medium-carbon, high-manganese steel with excellent abrasion-resistant properties and better workability than carbon steel of the same hardness.

ALUMINUM

Aluminum is a lightweight metal that has excellent corrosion resistance and electrical conductivity. It weighs approximately 28% as much as carbon steel and possesses a high strength-to-weight ratio.

ALUMINUM BRONZE

Aluminum bronze is a common name for a series of copper alloys containing 5 to 12% aluminum. Aluminum bronze alloys containing approximately 10% aluminum are popular because of their good mechanical properties.

BABBITT

This is a commercial name of a group of alloys used primarily as bearing materials. These materials may be tin or lead based or alloyed with antimony, arsenic, or copper.

BRASS

Brass is an alloy of copper and zinc, containing up to 40% zinc, with small amounts of other metals added to obtain increased strength, hardness, or corrosion resistance.

BRONZE

The term *bronze* is generally applied to any copper alloy that has a metal other than zinc or nickel as the principal alloying element. Some brasses are called bronzes because of their color or because they contain tin.

BUNA N

Buna N is a nitrile elastomer, or NBR rubber, known for its outstanding resistance to oil and fuel.

CARBON STEEL

Carbon steel's major properties depend on its 0.1 to 2% carbon content, without substantial amounts of other alloying elements.

CARTRIDGE BRASS

This is a commonly used copper alloy with a content of approximately 70% copper and 30% zinc.

CAST IRON

Cast iron is the metallic product that is obtained by reducing iron ore with carbon at a temperature that is high enough to render the metal fluid and cast it in a mold. Also called pig iron, it is generally hard and brittle and is neither malleable or ductile.

COMMERCIAL BRONZE

Commerical bronze is a copper alloy that contains approximately 90% copper and 10% zinc. Although commonly called bronze, it is actually a brass.

COPPER

Copper is a widely used metallic element. Its corrosion resistance, strength, fatigue resistance, formability, and electrical conductivity make it and its alloys well-suited for a variety of applications.

CUPRO-NICKEL

This is a copper alloy containing 10% to 30% nickel, which has excellent corrosion and stress corrosion cracking resistance.

GRAY IRON

Gray iron is cast metal widely used in sprockets and machinery parts, which consists essentially of iron and carbon, with a relatively large portion of its carbon in the form of graphite.

MALLEABLE IRON

This is a ferrous material that consists primarily of temper carbon and soft iron and includes small amounts of silicon, manganese, phosphorus, and sulfur. It offers strength, ductility, and machinability at a relatively low cost.

MONEL

An alloy of nickel and approximately 30% copper, monel has good mechanical and corrosion-resistant properties. K monel, a precipitation hardened of monel, has high resistance to impact and vibrational stresses.

NEOPRENE

Neoprene is a synthetic rubber, or elastomer, that is chemically, physically, and structurally similar to rubber.

POLYETHYLENE, UHMW

This is an ultrahigh-molecular-weight polyethylene, a self-lubricating plastic with low coefficient of friction and excellent abrasion and impact resistance.

POLYURETHANE

Polyurethane is a copolymer characterized by good tensile strength, tear strength, and abrasion resistance. Polyurethane elastomers possess the rigidity of plastics and the resilience of rubber. They can be formulated for injection molding or casting.

RED BRASS

A copper alloy with good cold working properties and superior corrosion resistance, red brass contains approximately 85% copper and 15% zinc.

SILICON BRONZE

Silicon bronze is a bronze that contains small amounts of silicon and manganese. It has greater strength and weldability than copper with comparable corrosion resistance.

STELLITE

This is a hard, wear- and corrosion-resistant family of nonferrous alloys containing cobalt (20–65%), chromium (11–32%), and tungsten (2–5%).

TOOL STEEL

Tool steel contains a high carbon and alloy content, characterized by high hardness and resistance to abrasion.

WHITE IRON

White iron is an iron containing 2 to 6% carbon present as iron carbides. White iron cannot be worked, but it has good casting properties and has applications where a cheap material where a cheap material with hardness and wear resistance is required.

WROUGHT IRON

Wrought iron is one of the purest grades of commercial iron, containing approximately 0.1% carbon and very small percentages of other nonmetals. It is malleable, ductile, and flexible and can be forged and welded.

EFFECTS OF ALLOYING ELEMENTS

ALUMINUM (Al)

Aluminum is a common deoxidizer used in the manufacture of steel to reduce the oxygen content and prevent a reaction between carbon and oxygen during solidification. It is also used to control grain size.

BORON (B)

Boron can be added to steel in amounts of 0.0005 to 0.003% to improve hardenability. When used in combination with other alloying elements, boron can help increase the depth of hardening during quenching.

CARBON (C)

Steel owes it properties chiefly to various amounts of carbon. As the amount of carbon increases up to 0.8 or 0.9%, the metal becomes harder and possesses greater tensile strength and becomes more responsive to heat treatment.

CHROMIUM (Cr)

Chromium is added to alloy steel in amounts of up to 1.5% to increase hardenability. If added in amounts greater than 4%, chromium confers an ability to resist corrosion. Stainless steels generally contain 12 to 30% chromium.

COBALT (Co)

Cobalt increases the strength and hardness of steels and allows the use of higher quenching temperatures. It is also used to intensify the effects of other major elements in more complex alloys.

COLUMBIUM (Nb)

The use of columbium in 18-8 stainless reduces harmful carbide precipitation and resultant inter-granular corrosion. Welding electrodes containing colombian are used in welding both titanium and columbium bearing stainless steels since titanium would be lost in the weld arc, whereas columbium is carried over into the weld deposits.

COPPER (Cu)

Copper is normally added in amounts of 0.15 to 0.25% to improve resistance to atmospheric corrosion and to increase tensile and yield strengths with only a slight loss in ductility. Higher strength properties can be obtained by precipitation hardening copper-bearing steel.

IRON (Fe)

Iron is the chief element of steel. Commercial iron usually contains other elements present in varying quantities, which produce the required mechanical properties. Iron lacks strength, is very ductile and soft, and does not respond to heat treatment to any appreciable degree. It can be hardened somewhat by cold working but not nearly as much as even a plain low carbon steel.

LEAD (Pb)

Lead is used to improve the machinability of steel. When the lead is finely divided and uniformly distributed, it has no known affect on the mechanical properties of the steel in the strength levels most commonly specified. It is usually added in amounts from .015% to 0.35%.

MANGANESE (Mn)

Manganese is an important alloying element present in all steel. It functions as a deoxidizer and degasifier, and is also used to impart strength and responsiveness to heat treatment. Manganese is usually present in quantities from 0.5% to 2%, although some specialty steels may contain as much as 10% to 15% manganese.

MOLYBDENUM (Mo)

Molybdenum increases strength, hardness, hardenability, and toughness. Molybdenum improves machinability and resistance to corrosion and is an important means of assuring high creep strength. It is generally used in comparatively small quantities ranging from 0.10 to 0.40%, and it intensifies the effects of other alloying elements.

NICKEL (Ni)

Nickel increases strength and toughness. Steels containing nickel usually have more impact resistance at low temperatures. Certain stainless steels employ nickel up to about 20%.

PHOSPHORUS (P)

Some phosphorus is present in all steel. It increases yield strength, hardness, and improves machinability.

SILICON (Si)

Silicon is one of the most common deoxidizers and degasifiers used during steel manufacture. It also may be present in various quantities up to 1% in the finished steel and has a beneficial effect on certain properties such as tensile and yield strength. It is also used in special steels to improve the hardenability and forgeability.

SULPHUR (S)

Sulphur increases machinability in free cutting steels. Sulphur is detrimental to the hot forming properties.

TITANIUM (Ti)

Titanium is a stabilizing element added to 18-8 stainless steels to make them immune to harmful carbide precipitation. It is sometimes added to low carbon sheets to make them more suitable to porcelain enameling.

TUNGSTEN (W)

Tungsten is used as an alloying element in tool steel and tends to produce a fine, dense grain and keen cutting edge when used in relatively small quantities. When used in quantities of 17 to 20%, and in combination with other alloys, it produces a high speed steel, which retains its hardness at the high temperatures developed in high-speed cutting. It is usually used in combination with chrome or other alloying elements.

VANADIUM (V)

Vanadium, usually in quantities from 0.15 to 0.20%, retards grain growth, permitting higher quench temperatures. It also intensifies the effects of other alloying elements.

Abbreviations and Acronyms

ABS	Acrylonitrile-butadiene-styrene
AC	Alternating current
ACI	Alloy Casting Institute
AFS	American Foundryman's Society
AFBMA	Anti-Friction Bearing Manufacturers Association
AGMA	American Gear Manufacturers Association
AIChE	American Institute of Chemical Engineers
AISC	American Institute of Steel Construction
AISI	American Iron and Steel Institute
ANSI	American National Standard Institute
API	American Petroleum Institute
APWA	American Public Works Association
AR	Abrasion-resistant
ASA	American Standards Association
ASCE	American Society of Civil Engineers
ASME	American Society of Mechanical Engineers
ASTM	American Society for Testing and Materials
AWG	American Wire Gauge
AWS	American Welding Society
AWWA	American Water Works Association
BAT	Best Available Technology
bbl	Barrel
BG	Birmingham gauge
BHN	Brinell hardness number
BHP	Brake horsepower
BIPM	International Bureau of Weights and Measures
BLS	Bureau of Labor Statistics
BOD	Biochemical oxygen demand
BSI	British Standards Institute
B&S	Brown & Sharpe
Btu	British thermal unit
BWG	Birmingham wire gauge
BWR	Boiling water reactor

CAD	Computer aided design
CAM	Computer aided manufacture
CDS	Cold drawn steel
CEMA	Conveyor Equipment Manufacturers Association
C&F	Cost and freight
CFR	Code of federal regulations
cfs	Cubic feet per second
CI	Cast iron
CIF	Cost, insurance and freight
CPM	Critical path method
CPVC	Chlorinated polyvinyl chloride
CRS	Cold rolled steel
CSA	Canadian Standards Association
CSO	Combined sewer overflow
DC	Direct current
DFT	Dry film thickness
DI	Deionization
DI	Ductile iron
DIN	Deutsche Industrie Normal
DO	Dissolved oxygen
DOE	Department of Energy
DP	Differential pressure
DP	Dimetral pitch
DWF	Dry weather flow
EDR	Electrodialysis reversal
EEI	Edison Electric Institute
EPA	Environmental Protection Agency
EPRI	Electric Power Research Institute
ERDA	Energy Research and Development Administration
FAS	Free alongside ship
FDA	Food and Drug Administration
FHCS	Flat head cap screw
FmHA	Farmers Home Administration
FOB	Free on board
fpm	Feet per minute
fps	Feet per second
FRP	Fiberglass reinforced plastic
FWPCA	Federal Water Pollution Control Act
gpm	Gallons per minute
G&A	General & administration
HDPE	High-density polyethylene
HHCS	Hex head cap screws
HOA	Hand-off-automatic
HRS	Hot rolled steel
HT	Heat treat
HWL	High water level
I/A	Innovative and alternative

ICC	Interstate Commerce Commission
ICC	International Chamber of Commerce
ICE	Institute of Civil Engineers
ID	Inside diameter
IEC	International Electrotechnical Commission
IEEE	Institute of Electrical & Electronic Engineers
IES	Illuminating Engineers Society of North America
IFI	International Fasteners Institute
I/O	Input/output
ISO	International Standards Organization
IX	Ion exchange
JIC	Joint Industrial Council
LC	Letter of credit
LPG	Liquified petroleum gas
LWL	Low water level
MED	Multiple effect distillation
mgd	Million gallons per day
mg/l	Milligrams per liter
MIL	Military specification
mks	Meter-kilogram-second
MPN	Most probable number
MSF	Multi-stage flash evaporation
MUD	Municipal utility district
MUS	Minimum ultimate strength
mW	Megawatt
NACE	National Association of Corrosion Engineers
NACM	National Association of Chain Manufacturers
NBS	National Bureau of Standards
NEC	National Electrical Code
NEMA	National Electrical Manufacturers Association
NFPA	National Fire Protection Association
NPDES	National Pollution Discharge Elimination System
NPSH	Net Positive Suction Head
NPT	National Pipe Thread
NRDC	National Resources Defense Act
NSF	National Science Foundation
NTIS	National Technical Information Service
OD	Outside diameter
OEM	Original equipment manufacturer
OHL	Overhung load
O&M	Operation and maintenance
ORNL	Oak Ridge National Laboratory
OSHA	Occupational Safety and Health Act
OSW	Office of Saline Water
OTS	Office of Technical Services
PC	Personal computer
PD	Pitch diameter

PE	Professional engineer
PERT	Program evaluation review technique
P&ID	Process and instrumentation diagram
PIV	Positive infinitely variable
PH	Precipitation hardened
PHD	Peak hourly demand
PLC	Programmable logic controller
POTW	Publically owned treatment works
ppm	Parts per million
psi	Pounds per square inch
psia	Pounds per square inch, absolute
psig	Pounds per square inch, gauge
PUC	Public Utilities Commission
PVC	Polyvinyl chloride
PWR	Pressurized water reactor
Q	Flow
QA	Quality assurance
QC	Quality control
RACT	Reasonably available control technology
RAS	Return activated sludge
RBC	Rotating biologial contactor
RCRA	Resource Conservation and Recovery Act
R&D	Research and development
RFP	Request for proposal
RFQ	Request for quotation
RMA	Rubber Manufacturers Association
rpm	Revolutions per minute
RO	Reverse osmosis
SAE	Society of Automotive Engineers
SCR	Silicon controlled rectifier
SDI	Silt density index
SDWA	Safe Drinking Water Act
SF	Service factor
SHCS	Socket head cap screw
SI	Systéme international
SIC	Standard industrial classification
SME	Society of Manufacturing Engineers
SS	Suspended solids
SSPC	Steel Structures Painting Council
SST	Stainless steel
SVI	Sludge volume index
SWG	Standard (British) wire gauge
TBE	Thread both ends
TDH	Total dynamic head
TEFC	Totally enclosed fan cooled
TOE	Thread one end
TDS	Total dissolved solids

TS	Total solids
TSS	Total suspended solids
TWL	Top water level
TWS	Traveling water screen
UF	Ultrafiltration
UHMW	Ultrahigh-molecular-weight
UL	Underwriters Laboratories
UNS	Unified numbering system
USDA	U.S. Department of Agriculture
USGS	U.S. Geological Survey
USPHS	U.S. Public Health Service
USSG	United States Standard Gauge
UV	Ultraviolet
V	Volt, velocity, or volume
VFD	Variable frequency drive
VHN	Vickers hardness number
WHO	World Health Organization
W&M	Washburn & Moen
WPCF	Water Pollution Control Federation
WPCP	Water Pollution Control Plant
WQA	Water Quality Act of 1987
WS	Welded steel
WTP	Water treatment plant
WWEMA	Water and Wastewater Equipment Manufacturers Association
WWTP	Wastewater treatment plant
XP	Explosion proof

Conversion Tables

TO CONVERT:	MULTIPLY BY:	TO OBTAIN:
acre-feet	43,560	cubic feet
acre-feet	325,900	gallons
bars	14.5	pounds/square inch
centimeters	0.003281	feet
centimeters	0.3937	inches
centimeters	10	millimeters
centimeters	10,000	microns
centimeters/second	1.969	feet/minute
centimeters/second	0.03281	feet/second
centimeters/second	0.6	meters/minute
centimeters/second	0.2237	miles/hour
centimeters/second	0.036	kilometers/hour
cubic centimeters	0.06102	cubic inches
cubic centimeters	1,000,000	cubic meters
cubic centimeters	26,420	gallons
cubic feet	1728	cubic inches
cubic feet	0.02832	cubic meters
cubic feet	0.03704	cubic yards
cubic feet	7.48052	gallons (U.S. liquid)
cubic feet	28.32	liters
cubic feet/minute	0.1247	gallons/second
cubic feet/minute	0.4702	liters/second
cubic feet/minute	62.43	pounds water/minute
cubic feet/second	0.646317	million gallons/day
cubic feet/second	448.831	gallons/minute
cubic meters	35.31	cubic feet
cubic meters	61,023	cubic inches
cubic meters	1.308	cubic yards
cubic meters	264.2	gallons (U.S. liquid)

TO CONVERT:	MULTIPLY BY:	TO OBTAIN:
cubic yards	27	cubic feet
cubic yards	0.7646	cubic meters
cubic yards	202	gallons (U.S. liquid)
cubic yards/minute	0.45	cubic feet/second
cubic yards/minutes	3.367	gallons/second
days	86,400	seconds
days	1440	minutes
degrees (angle)	0.01745	radians
degrees/second	0.01745	radians/second
degrees/second	0.1667	revolutions/minute
feet	30.48	centimeters
feet	0.3048	meters
feet	304.8	millimeters
feet	12,000	mils
feet of water	0.03048	kilograms/sq centimeter
feet of water	304.8	kilograms/sq meter
feet of water	62.43	pounds/sq foot
feet of water	0.4335	pounds/sq inch
feet/minute	0.5080	centimeters/second
feet/minute	0.01667	feet/second
feet/minute	0.3048	meters/minute
feet/second	30.48	centimeters/second
feet/second	18.29	meters/minute
foot-pounds	0.1383	kilogram-meters
foot-pounds/minute	0.01667	foot-pounds/second
foot-pounds/minute	0.0000303	horsepower
foot-pounds/minute	0.0000226	kilowatts
foot-pounds/second	4.6263	btu/hour
foot-pounds/second	0.07717	btu/minute
foot-pounds/second	0.001818	horsepower
foot-pounds/second	0.001356	kilowatts
gallons	3785	cubic centimeters
gallons	0.1337	cubic feet
gallons	231	cubic inches
gallons	0.003785	cubic meters
gallons	0.004951	cubic yards
gallons	3.785	liters
gallons (U.S.)	0.83267	gallons (imperial)
gallons of water	8.337	pounds of water
gallons/minute	0.002228	cubic feet/second
gallons/minute	0.06308	liters/second
gallons/minute	8.0208	cubic feet/hour

TO CONVERT:	MULTIPLY BY:	TO OBTAIN:
grams	0.001	kilogram
grams	1000	milligrams
grams	0.00205	pounds
grams/centimeter	0.000056	pounds/inch
grams/cu centimeter	62.43	pounds/cubic foot
grams/cu centimeter	0.03613	pounds/cubic inch
grams/liter	8345	pounds/1000 gallons
grams/liter	0.062427	pounds/cubic foot
grams/sq centimeter	2.0481	pounds/square foot
horsepower	42.44	btu/minute
horsepower	33,000	foot-pounds/minute
horsepower	550	foot-pounds/second
horsepower	0.7457	kilowatts
horsepower	745.7	watts
horsepower-hours	1,980,000	foot-pounds
horsepower-hours	0.7457	kilowatt-hours
inches	2.54	centimeters
inches	0.0254	meters
inches	25,400	microns
inches	25.4	millimeters
inches	1000	mils
kilograms	1000	grams
kilograms	2.2046	pounds
kilograms	0.0009842	tons (long)
kilograms	0.001102	tons (short)
kilograms/cu meter	0.06243	pounds/cubic foot
kilograms/cu meter	0.00003613	pounds/cubic inch
kilograms/meter	0.672	pounds/foot
kilograms/sq cm	2048	pounds/square foot
kilograms/sq cm	14.22	pounds/square inch
kilograms/sq meter	0.2048	pounds/square foot
kilograms/sq meter	0.001422	pounds/square inch
kilometers/hour	27.78	centimeters/second
kilometers/hour	54.68	feet/minute
kilometers/hour	0.9113	feet/second
kilometers/hour	16.67	meters/minute
kilometers/hour	0.6214	miles/hour
kilopascals	0.145018	pounds/square inch
kilowatts	1.341	horsepower
kilowatts	73.76	foot-pounds/second
kilowatt-hours	2,655,000	foot-pounds
kilowatt-hours	1.341	horsepower-hours

TO CONVERT:	MULTIPLY BY:	TO OBTAIN:
liters	1000	cubic centimeters
liters	0.03531	cubic feet
liters	61.02	cubic inches
liters	0.001	cubic meters
liters	0.2642	gallons (U.S. liquid)
liters/minute	0.0005886	cubic feet/second
liters/minute	0.004403	gallons/minute
$\log_{10} n$	2.303	$\ln n$
$\ln n$	0.4343	$\log_{10} n$
meters	100	centimeters
meters	3.281	feet
meters	39.37	inches
meters	1000	millimeters
meters/minute	1.667	centimeters/second
meters/minute	3.281	feet/minute
meters/minute	0.05468	feet/second
meters/minute	0.03728	miles/hour
meters/second	196.8	feet/minute
meters/second	3.281	feet/second
meters/second	3.6	kilometers/hour
meters/second	2.237	miles/hour
meter-kilogram	7.233	pound-feet
microns	1,000,000	meters
microns	0.0000394	inches
miles	1.609	kilometers
miles	5280	feet
miles/hour	44.7	centimeters/second
miles/hour	88	feet/minute
miles/hour	1.467	feet/second
miles/hour	1.6093	kilometer/hour
miles/hour	26.82	meters/minute
miles/hour	0.01667	miles/minute
milligrams	0.001	grams
milligrams⅓liter	1.0	parts/million
milliliters	0.001	liters
millimeters	0.1	centimeters
millimeters	0.003281	feet
millimeters	0.03937	inches
millimeters	0.001	meters
millimeters	39.37	mils
million gallons/day	1.54723	cubic feet/second
mils	0.00254	centimeters

TO CONVERT:	MULTIPLY BY:	TO OBTAIN:
mils	0.001	inches
parts/million	1	milligrams/liter
parts/million	8.345	pounds/million gallons
pounds of water	0.01602	cubic feet
pounds of water	27.68	cubic inches
pounds of water	0.1198	gallons
pounds-feet	0.1383	meters-kilograms
pounds/cubic feet	16.02	kilograms/cubic meter
pounds/cubic feet	0.0005787	pounds/cubic feet
pounds/cubic inch	1728	pounds/cubic foot
pounds/feet	1.488	kilograms/meter
pounds/square inch	2.307	feet of water
pounds/square inch	703.1	kilogram/square meter
pounds/square inch	144	pounds/square foot
pounds/square inch	0.0703	kilograms/sq centimeter
pounds/square inch	6.8957	kilopascals
pounds/square foot	0.01602	feet of water
pounds/square foot	4.882	kilograms/squre meter
pounds/square foot	0.006944	pounds/square inch
radians	57.296	degrees
radians/second	9.549	revolutions/minute
revolution/minute	6	degrees/second
square centimeters	0.001076	square feet
square centimeters	0.1550	square inches
square centimeters	0.0001	square meters
square feet	0.00002296	acres
square feet	929	square centimeters
square feet	144	square inches
square feet	0.0929	square meters
square feet	0.1111	square yards
square inches	6.452	square centimeters
square inches	0.006944	square feet
square inches	645.2	square millimeters
square meters	0.0002471	acres
square meters	10000	square centimeters
square meters	10.76	square feet
square meters	1550	square inches
square meters	1.196	square yards
square millimeters	0.00155	square inches
square yards	9	square feet
square yards	0.8361	square meters
tons (long)	1016	kilograms

TO CONVERT:	MULTIPLY BY:	TO OBTAIN:
tons (long)	2240	pounds
tons (long)	1.12	tons (short)
tons (metric)	1000	kilograms
tons (metric)	2205	pounds
tons (short)	907.18	kilograms
tons (short)	2000	pounds
tons (short)	0.89287	tons (long)
tons (short)	0.9078	tons (metric)
tons of water/24 hrs	83.33	pounds of water/hour
tons of water/24 hrs	0.16643	gallons/minute
tons of water/24 hrs	1.3349	cubic feet/hour

Glossary

THIS glossary contains terms and product names frequently used within the water and wastewater screening industry. Some product names may be copyrighted, and some may be obsolete or no longer offered for sale.

Many products are licensed for sale by different suppliers in various countries or geographical regions. When possible, these products include the names of their suppliers with their respective areas of responsibility.

Acme Former screening equipment manufacturer.

Agisac Sock-type screening sack by Hydro-Aerobics (US) and Copa Group (UK).

Alligatoring A surface defect common to coal-tar paints caused by uneven hardening of paint film, stimulated by the sun's rays. Results in the upper layer of paint's surface cracking and slipping over softer lower stratum.

Alloy Material with metallic properties composed of two or more elements, of which at least one is a metal.

Annealing The heating and controlled cooling of a solid material to reduce hardness, improve machinability or obtain desired mechanical, physical, or other properties.

Anode The positive electrode where current leaves the solution.

Approach Velocity The average water velocity of fluid in a channel upstream of a screen or other obstruction.

Aqua-guard In-channel filter screen by Parkson Corp. (USA), Noggerath (Germany), and Andritz Sprout-Bauer (France).

Aquifer A geological formation containing a large quantity of water.

Ashbrook-Simon-Hartley Manufacturer formerly offering a bar screen.

Auto-Rake Hydraulically driven, reciprocating rake bar screen by Franklin Miller Co.

Auto-Retreat Automatic bar screen control system by Infilco Degremont.

B-10 Life The "rated life" that defines the number of revolutions that 90% of a group of identical bearings will complete or exceed before first evidence of fatigue develops.

Babbitted Bearing A bearing using babbitt, a tin- or lead-based alloy, as an antifriction lining.

Backlash Movement (if any) of chain along the pitch line of the sprocket when the direction of chain travel is reversed.

Bandscreen Common term (European) for a traveling water screen. Also "traveling band screen."

Barry Rake Trash rake by Cross Machine, Inc.

Barminutor Combination bar screen and comminuting device by Yeomans Chicago Corp.

Basket The individual screening elements used in a traveling water screen consisting of a wire mesh panel and its structural frame. Also called a "tray."

Bauer Former name of Andritz Sprout-Bauer.

Berry Hydro Rake Trash rake by Cross Machine, Inc.

Bioflush Trash rake by E. Beaudrey & Co.

Bitumastic Bituminous coating by Carboline Co.

Blakeborough Former equipment manufacturer acquired by Brackett-Green, Ltd.

BOD Biochemical oxygen demand. A standard measure of wastewater strength that quantifies the mg/L of oxygen consumed in a stated period of time, usually five days.

Boot The lower, or bottom, portion of a screen structure.

Bosker Trash rack cleaner by Brackett Green.

Brinell Hardness A hardness rating determined by the surface area of the cross section of an indentation made by a 1-cm steel ball pressed into a sample material with a standard force.

Brownie Buster Device used to break up biodegradable solids from wastewater plant influent by Enviro-Care.

Bubbler System Common terminology for pneumatic-type differential level controller.

Bulkhead A partition of wood, rock, concrete, or steel used for protection from water.

Bushings, Chain The bearing surface for pin rotation when chain articulates over a sprocket. Bushings also provide the bearing surface for chain rollers or sprocket contact in rollerless chain.

Cable cylinder A hydraulic or pneumatic cylinder that utilizes a cable and pulley(s) system.

Caisson Watertight structure used for underwater work.

Caldicot In-channel filter screen by Advance Wastewater Treatment, Ltd.

Calender The process of passing wire cloth through a pair of heavy rollers to reduce its thickness or flatten it to produce a smooth surface.

Carbon Steel Steel whose major properties depend on its carbon content, without substantial amounts of other alloying elements.

Carburizing Surface heat treatment process done primarily to increase wear resistance of sprockets or chain pins, rollers, and bushings.

Catenary A curve formed by a chain hanging freely from two points not in the same vertical line.

Cathode The negative electrode where the current leaves the solution.

Cathodic Protection The use of an impressed current to prevent or to reduce the rate of corrosion of a metal in an electrolyte by making the metal the cathode for the impressed current.

Cavitation The formation of gas- or vapor-filled cavities within a liquid, formed by mechanical forces.

CE-Bauer Former name of Andritz Sprout-Bauer.

Centrifugal Casting A casting process usually used for tubular products where a weighed portion of molten metal is poured into a metal mold while it is being spun on a horizontal-spindle machine.

Chabelco Brand name of Rexnord, Inc. chain products.

Chill A metal or graphic insert embedded in the surface of a sand mold or core or placed in a mold cavity to increase the cooling rate and produce a wear-resistant surface.

Chordal Action The effect produced by the center of a chain joint being forced to follow arcs instead of chords of the sprocket pitch circle.

Clarifier A quiescent tank or basin that is used to remove suspended solids through gravity settling.

Climber Reciprocating rake bar screen by Infilco Degremont Inc.

ClimbeRack Nonlubricated pin rack for reciprocating rake bar screen by Infilco Degremont Inc.

Coanda Effect The tendency of a liquid coming out of a nozzle or orifice to travel close to the wall contour even if the wall's direction of curvature is away from the jet's axis.

Cofferdam A temporary dam, usually of sheet piling, built to provide access to an area that is normally submerged.

Cog Rake Reciprocating rake bar screen by FMC Corp.

Cogwheel A wheel with teeth around its edge.

Colloid Suspended solids with a diameter less than 1 micron that cannot be removed by sedimentation alone.

Combi-Guard Packaged screening unit by Andritz Sprout-Bauer S.A.

Combination Chain Popular chain for heavy-duty conveyor and elevator applications. This chain has cast block links with steel pins and connecting bars.

Combined Sewer Overflow (CSO) Wastewater flow that consists of storm water and sanitary sewage.

Comminutor A circular screen with cutters that grind large sewage solids into smaller setteable particles.

Composite A material composed of two materials bonded together, with one serving as a matrix surrounding the particles or fibers of the other.

Compound 146 Polyurethane sprocket tooth insert material by FMC MHS's Division.

Cone Screen Internally fed rotary fine screen by Andritz Sprout-Bauer (Western Hemisphere) and Contra-Shear Engineering, Ltd.

Conoscreen Rotating disc microscreen by Purator Waagner-Biro.

Cont-flo Reciprocating, back rake bar screen by John Meunier, Inc.

Contra-Shear Screening equipment product line by Andritz Sprout-Bauer (Western Hemisphere) and Contra-Shear Engineering, Ltd.

Copasacs Fine screening sack by Hydro-Aerobics (US) and Copa Group (UK).

Copa Screen Packaged screening plant with bar screen, macerator, and screenings dewatering unit by Longwood Engineerings Co.

Copasocks Sock-type screening sack by Hydro-Aerobics (US) and Copa Group (UK).

Copatrawl Sock-type screening sack by Hydro-Aerobics (US) and Copa Group (UK).

Coping The top or covering of an exterior masonry wall.

Corten High-strength, low-alloy steel with enhanced atmospheric corrosion resistance by US Steel Corp.

CPC-Microfloc Former name of CPC Products division of Wheelabrator Engineered Systems, Inc.

Curtain Wall An external wall that is not load bearing. Usually refers to a wall that extends down below the surface of the water to prevent floating objects from entering a screen forebay. Also called a "skimmer wall."

Davy Bamag Former name of Bamag GmbH.

Delrin High-molecular-weight acetal resin polymer material by E. I. du Pont de Nemours & Co.

Delt△ Traveling water screen chain by Envirex, Inc.

Detritus, Inorganic A heterogeneous mass of fragments of stone.

Detritus, Organic Decaying organic matter such as root hairs, stems, and leaves usually found on the bottom of a water body.

Diametral Pitch The ratio of the number of teeth in a gear to the diameter of the pitch circle.

Diamond Gate Screenings press by Andritz Sprout-Bauer (Western Hemisphere) and Contra-Shear Engineering, Ltd.

Dijbo Hydraulically operated trash rake by Landustrie Sneek BV.

Dimminutor   Comminutor by Franklin Miller, Inc.

Discostrainer   Rotary fine screening device by Hycor Corp.

Discreen   Sewage screen by Dresser Pump, Waste-Tec Operations.

Disposable Water Systems   Former name of JWC Environmental.

Doctor Blade   A scraping device used to remove or regulate the amount of material on a belt, roller, or other moving or rotating surface.

Dokwed   Back cleaned bar screen by Hubert Stavoren BV.

Domestic Wastewater   Wastewater originating from sanitary conveniences in residential dwellings, office buildings, and institutions.

Dorrco   Arc-type screen by Dorr-Oliver.

Double Crimp   A type of wire mesh with corrugations in both the warp and shute wire to lock the wires in position.

Dresser/Jeffrey   Company whose screening equipment product line was acquired by Jones + Atwood.

Drip Proof   Designation for motor enclosure with ventilating openings constructed so that drops of liquids or solids falling on the motor at an angle of fifteen degrees or less will not enter the unit either directly or by running along an inwardly inclined surface.

Drumshear   Rotating fine screen by Aer-O-Flo Environmental, Inc.

Durometer Hardness   The hardness of a material measured with an instrument utilizing a small drill or blunt indenter point under pressure.

Duroy   Roller chain material combination by Envirex, Inc.

Dutch Weave   Wire mesh woven similar to "plain weave," except that warp wires are usually larger than shute wires, and shute wires are closely spaced, resulting in a dense weave.

Dyna-Grind   Screenings grinder by FMC-Material Handling Systems Division.

Dynasieve   Externally fed rotary fine screen by Andritz Sprout-Bauer.

Effluent   Partially or completely treated water or wastewater flowing out of a basin or treatment plant.

Elastic Limit   The greatest unit stress that a material is capable of withstanding without permanent deformation.

Elastomers   Synthetic rubbers of hydrocarbon and polymeric materials similar in structure to plastic resins.

Elbow Rake   Hydraulic trash rake by former manufacturer Acme Engineering Corp.

Electrolyte   A substance that dissociates into two or more ions when it is dissolves in water.

Embrittlement   Reduction in ductility of materials due to exposure to certain environments or temperatures.

Entrainment   The incorporation of small organisms, including the eggs and larvae of fish and shellfish, into an intake system.

Enviropress   Piston-type screenings press by Environmental Engineering Ltd.

Estuary   The mouth of a river in which the river's current meets the sea's tide.

Expanded Metal   An open metal network produced by stamping or perforating sheet metal.

Explosion Proof   Designation for a motor or electrical enclosure designed to withstand a gas or vapor explosion within the unit and to prevent ignition of gas or vapor surrounding the unit by sparks, flashes, or explosions that may occur within the unit.

FBS   Fine bar screen by Jones + Atwood.

Fecascrew   Screenings screw press by Hydropress Wallender & Co.

Fecawash   Screenings washing and conveying unit by Hydropress Wallender & Co.

Feedwater   Water that, in the best practice, is demineralized and heated to nearly boiler temperature and deaerated before being pumped into a steam boiler.

Filtester   Test apparatus used to predict filterability of microscreen fabric by Weir Pumps, Ltd.

Fish Ladder   A structure that permits fish to bypass a dam. The ladder resembles a staircase alongside the dam spillway. Water flows down the staircase, with each step built high enough to allow the fish to jump successive steps, against the flow.

Flex-A-Gard   Polyurethane bar screen products by Flex-A-Seal.

Flint Rim   Tradename of FMC/PTC chill rim cast sprocket.

Flo-Conveyor   Horizontal screenings conveyor by Enviro-Care Co.

Flo-Lift   Vertical screenings lift device by Enviro-Care Co.

Flo-Screen   Screw-operated reciprocating bar screen by Enviro-Care Co.

Flowminutor   Comminutor by Enviro-Care Co.

Flux   The amount of some quantity flowing across a given area per unit time.

Forebay   A reservoir at the end of a pipeline or channel.

Forging   Plastically deforming metal, usually hot, into desired shapes with compressive force.

Frazil Ice   Granular or spike-shaped ice crystals that form in supercooled water that is too turbulent to permit coagulation into sheet ice.

Free-Slide   Traveling water screen wire mesh and tray configuration by Envirex, Inc.

Frontloader   Reciprocating rake bar screen by Schreiber Corp.

Frontrunner   Reciprocating rake bar screen by Jones + Attwood.

Fry   Juvenile fish.

Galling   Development of a condition on the live bearing surface of mating parts where excessive friction results in localized welding and a further roughening of contact surfaces.

Galvanic Couple   The connection of two dissimilar metals in an electrolyte that results in current flow through the circuit.

Galvanize   An electrolytic or hot dripping process to coat steel products with a coating of zinc.

Geiger Screen   Reciprocating rake bar screen named after its developer/manufacturer, also marketed as a "Climber" screen.

Girasieve   Externally fed rotary fine screen by Andritz Sprout-Bauer.

Grabber   Reciprocating rake bar screen by Hycor, Corp.

Grit   The dense, mineral, suspended matter present in a water or wastewater, such as sand, silt, or cinders.

Grit Chamber   A settling chamber used to remove grit from organic solids through sedimentation or an air-induced spiral agitation.

Grout   Fluid, or semi-fluid, cement slurry for pouring into joints of brickwork or masonry.

Hawco   Former name of Screening Systems International.

Hazardous Area, Class 1   Locations where flammable gases or vapors may be present in the air in sufficient quantities to produce explosive or ignitable mixtures.

Headloss   The difference in water level between the upstream and downstream sides of a screen.

Helical Gear   Gear wheels running on parallel axes with teeth cut oblique to the gear axis.

Heliclean   Combination in-channel fine screen and conveying unit by Hycor, Corp.

Heli-Press   Screenings compactor by Vulcan Industries, Inc.

Helisieve   In-channel rotary fine screen by Hycor, Corp.

Helixpress   Spiral dewatering-conveying screenings press by Hycor Corp.

Hercules   Screening equipment division of Atlas Polar Co.

Hevi-Duty   Traveling water screen components by Envirex, Inc.

Holiday   Any discontinuity or bare spot in a coated surface.

Hunting Tooth Sprocket   A sprocket with two sets of effective teeth arranged so that each set makes contact with the chain with alternate revolutions of the wheel.

Hydra-Press   Hydraulic screenings compactor by Vulcan Industries.

Hydra-Press   Solids dewatering press by Dontech, Inc.

Hydrasieve   Static screen by Andritz Sprout-Bauer, Inc.

Hydroburst   Passive screen air backwash system by Wheelabrator Engineered Systems, Inc.

Hydro-Dri   Screenings press by Serpentix Conveyor Corp.

Hydroflush   Cable operated bar screen by Beaudrey Corp.

Hydrorake   Trash rake by Atlas Polar Co., Hercules Division.

Hydroscreen   Static screen by Hycor Corp.

Hydro-Shears   Internally fed rotary fine screen by Dontech, Inc.

Hydrosil   Static screen by Spirac Engineering.

Hypress   Screenings press by Hycor Corp.

IDS Drumshear   Rotating fine screen by Aer-O-Flo Environmental, Inc.

Impingement   The entrapment of fish and other marine life against screening media that results when organisms cannot escape the area in front of the screen because of the velocity of the intake stream.

Infiltration Gallery   A horizontal underground conduit of screens or porous material to collect percolating water. Often placed under a river bed.

Influent   Water or wastewater flowing into a basin or treatment plant.

Invert   The lowest point of the internal surface of a drain, sewer, or channel at any cross section.

Johnson Filter Co.   Former name of Johnson Products Division of Wheelabrator Engineered Systems, Inc.

Johnson Screen   Wedgewire screen media by Wheelabrator Engineered Systems, Inc.

Journal   That part of a shaft that is supported by and turns in a bearing.

Journal Bearing   A cylindrical bearing that supports a cylindrical rotating shaft.

Key   A precision-made bar used to transfer torque between sprockets or pulleys and the shafts on which they are mounted. Keys may be of straight or tapered design and furnished with a "gib-head" to facilitate removal.

Komprimat   Fish/screenings separation system by Geiger GmbH & Co.

L-10 Life   *See* B-10 LIFE.

Lakos   Self-cleaning pump intake screening by Claude Laval Corp.

Laminar Flow   A flow situation in which fluid moves in parallel layers, usually with a Reynolds number less than 2000.

Launder   Trough used to transport water.

Ledward & Beckett   Former equipment manufacturer acquired by Brackett-Green, Ltd.

Lift Screen   Reciprocating rake bar screen by Envirex, Inc.

Link Belt   Trade name of FMC Corporation.

Liquifuge   Internally fed rotary fine screen by Vulcan Industries, Inc.

Liqui-Strainer   Externally fed rotary fine screen by Vulcan Industries, Inc.

Longopac   Screenings bagging system by Spirac Engineering.

Macerate   Chop or tear.

Macrofouling   Clogging of a condenser tubesheet with debris and biogrowth.

Mather Platt   Former name of Wier Pumps, Ltd. screening equipment division.

McGinnes-Royce   Former screening equipment manufacturer whose product line was acquired by Envirex, Inc.

Meehanite The name of several cast irons, cast for a specific purpose such as strength or corrosion resistance, having different combinations of mechanical and engineering properties.

Mensch Reciprocating rake bar screen by Vulcan Industries.

Mesh (as a number) The number of openings per lineal inch, measured from the center of one wire or bar to a point 1″ distant.

Micrasieve Pressure-fed fine screen by Andritz Sprout-Bauer.

Micro-Matic Microscreen by Lyco, Inc.

Micro-Pi Pressure fed rotary screen by Andritz Sprout-Bauer (Western Hemisphere) and Contra-Shear Engineering, Ltd.

Micro-Sieve Microscreen formerly offered by Passavant Corp.

Mill Grind or crush.

Milliscreen Internally fed rotary fine screen by Andritz Sprout-Bauer (Western Hemisphere) and Contra-Shear Engineering, Ltd.

Modius In-channel fine screen by Pro-Ent, Inc.

Mohs Hardness A measure of hardness of a material, as determined by the width of a scratch depth made under a prescribed load.

Moment of Inertia The sum of products formed by multiplying the mass of each element of a figure by the square of its distance from a specified line.

Monkey Screen Reciprocating rake bar screen by Brackett Green.

Monomer The basic molecule of a synthetic resin or plastic.

Muffin Monster Sewage grinders by JWC Environmental.

Muncher Sewage grinder by Dresser Pump, Waste-Tec Operations.

Munchpump Sewage grinder/pump package by Dresser Pump, Waste-Tec Operations.

Münster Trash rake cleaning mechanism by Lanustrie Sneek BV.

Net-Waste Screw press by Diemme USA.

NoCling Screening media tray insert by E. Beaudrey.

No-well Platform mounted traveling water screen by FMC Corp.

Nylon Plastic compound that offers excellent load bearing capabilities, low frictional properties, and good chemical resistance.

Paint Filter Test Test to determine free water content of sludge sample.

Para Cone Internally fed rotary fine screen by Andritz Sprout-Bauer (Western Hemisphere) and Contra-Shear Engineering, Ltd.

Parshall Flume A venturi-type flume used to measure flow.

Passive Screen Intake screening device that does not employ mechanical cleaning.

Passivation The changing of a chemically active surface of a metal to a much less reactive state. Usually done to stainless steel by immersion in an acid bath.

Pathwinder Screenings conveyor by Serpentix Conveyor Corp.

Penstock A pipe that transports water to a turbine for the production of hydroelectric energy.

Pickling Preferential removal of oxide or mill scale from the surface of a metal by immersion in an acid or an alkaline solution.

Pins, Chain Chain pins connect chain links. They are locked in the sidebars so all relative rotation occurs between the pin and the bushing.

Pinion The smaller of a pair of gear wheels or the smallest wheel of a gear train.

Pintle Chain Chain extensively used for elevating and conveying, consisting of one-piece links cast with two offset sidebars and coupled with steel pins. Lugs prevent turning of the pins in the sidebars, insuring that articulation will occur between the pin and the cored cylinder.

Pitch The length of one link of chain measured from pin centerline.

Pitch Diameter The diameter of the pitch circle of a sprocket or gear.

Plain Weave Wire mesh with warp and shute wires passing over and under the next adjacent wire in both directions.

Planetary Gear Train An assembly of meshed gears consisting of a central gear, a ring gear, and one or more intermediate pinions supported on a revolving carrier.

Plastic Deformation Permanent change in shape or size of a solid body without fracture, resulting from the application of sustained stress beyond the elastic limit.

Polymer A compound of high-molecular weight derived by the recurring addition of similar molecules.

Posirake Reciprocating rake bar screen formerly offered by Passavant Corp.

Powermatic Reciprocating rake bar screen by Brackett Green.

Powrclean Brand name of Aerators, Inc. multi-raked bar screen, developed by the former Welles Products Corp.

Precipitation Hardening Heat treating process used for some alloys that involves a solution heat treatment and rapid quenching followed by a low-temperature precipitation or aging treatment.

Preliminary Treatment Treatment step such as including screening and grit removal to prepare wastewater influent for further treatment.

Pressveyor Hydraulic screenings press by Hycor Corp.

Primary Treatment Treatment steps, including sedimentation and fine screening to produce an effluent suitable for biological treatment.

Profiling The action of a bar screen cleaning rake as it travels around the "profile" of an obstruction.

Profile Wire Term used to describe specially shaped wire that is generally triangular or trapezoidal in cross section.

Promal High-strength pearlitic malleable iron chain material by FMC/PTC.

Primary Treatment The initial treatment of wastewater, usually by sedimentation and/or screening, to remove suspended solids.

Putrescible Organic matter in a state of decay or decomposition.

Rake-O-Matic Hydraulically operated, reciprocating rake bar screen, formerly offered by BIF Division of General Signal.

Ranney Intake Surface water intake system utilizing a passive screen/caisson arrangement by Ranney Division of Hydro Group, Inc.

Raw Edge An unselvaged edge on a piece of woven wire mesh.

Reacher Reciprocating rake bar screen by Schloss Engineered Equipment, Inc.

Red Rubber Cast urethane bar screen rake and rake tooth segments by Rubber Millers, Inc.

Retrofit The extensive modification/upgrading of an existing screen or its replacement with a different type of screen.

Rex Trademark of Rexnord, Inc.

Reynolds Number A nondimensional number that measures the state of turbulence in a fluid system. It is calculated as the ratio of inertia effect to viscous effect.

Ristropf Screen Traveling water screen equipped with a fish bucket collection and sluice spray discharge system named for one of its developers, J. D. Ristropf.

Robo Bar screen by Vulcan Industries.

Robo Rover Traversing bar screen by Vulcan Industries.

Robo Stat Stationary bar screen by Vulcan Industries.

Rockwell Hardness A measure of hardness of a material as determined by the depth of indentation made by a twelve-degree conical diamond under a prescribed load.

Roll-Dry Internally fed rotary fine screen by Schlueter Co.

Roller Chain Chain equipped with a roller that revolves over a stationary bushing.

Rollers, Chain Chain rollers reduce the coefficient of friction in roller chain by rolling rather than sliding and minimize sprocket scrubbing as chain enters and leaves a sprocket.

Rotafine Rotary fine screen by Jones + Attwood.

Rotamat Screening equipment product line by Lakeside Equipment Co. (US) and Hans Huber (Europe).

Rotarc Rotary bar screen by John Meunier, Inc.

Rotasieve Externally fed rotary fine screen by Jones + Attwood.

Rotobelt In-channel fine screen by Dontech, Inc.

Roto-Brush Externally fed rotary fine screen by Dontech, Inc.

RotoClean Screenings washer by Parkson Corp.

Rotoclear Microscreen formerly offered by Walker Process Co.

Roto-Drum Internally fed rotary fine screen by Dontech, Inc.

Rotofilter Externally fed rotary fine screen by Sepra Tech.

Roto-Guard Drum screen/thickener by Parkson Corp.

Rotopac   Screenings compactor by John Meunier, Inc.

Rotopass   Externally fed rotary fine screen by Passavant-Werke AG.

Roto-Press   Screenings compactor by Roto-Sieve AB.

RotoPress   Screenings compactor by Parkson Corp.

Rotoshear   Internally fed rotary fine screen by Hycor.

Roto-Sieve   Internally fed rotating fine screen by Roto-Sieve AB.

Rotostep   In-channel fine screen by Hycor Corp.

Rotostrainer   Externally fed rotary screen by Hycor Corp.

Royce   Former screening equipment manufacturer whose product line was acquired by Envirex, Inc.

Run Down Screen   Another term for a static screen.

Ryertex   Fabric/plastic laminate bushing material manufactured by J. T. Ryerson & Son, Inc.

Sacrificial Anode   A sacrificial piece of metal, usually zinc or magnesium, that is electrically connected to a more noble metal in an electrolyte. The anode goes into solution at a disproportionate, accelerated rate to protect the more noble metal from corrosion.

Sani-Sieve   Static screen by Dontech, Inc.

Sanitary Wastewater   Domestic wastewater without storm and surface runoff, which originates from sanitary conveniences.

Schoop Process   Process for coating steel that uses a blast of air to spray a mist of molten metal onto the surface to be protected.

Screezer   Combination screening and dewatering device by Jones + Attwood.

ScruPac   Screw-type screenings compactor by Vulcan Industries.

Secondary Treatment   The treatment of wastewater through biological oxidation after primary treatment.

Section Modulus   The ratio of the moment of inertia of the cross section of a beam undergoing flexure to the greatest distance of an element of the beam from the neutral axis.

Sedimentation   The removal of setteable suspended solids by gravity in a clarifier.

Selvage   A finished edge on wire mesh to prevent its unravelling.

Septage   Liquid and solid contents of a septic tank.

Service Factor   A multiplier that, when applied to the rated power, indicates the permissible power loading that may be carried under the conditions specified.

Shear Pin Sprocket   A drive sprocket equipped with a shear pin device used to protect equipment from jamming or overloads. The replaceable shear pins transmit the required torque under normal conditions and fail when overloaded.

Sherardizing   A process for protecting iron from corrosion by means of a corrosion-resistant layer of zinc on the iron surface.

Shore (Scleroscope) Hardness A measure of hardness by dropping a diamond-tipped "hammer" on a material and the rebound taken as an index of hardness.

Shute The horizontal wire in woven wire mesh, also called the "weft" wire.

Sidebars, Chain Chain sidebars are the tensile members of a chain.

Side Hill Screen Another term for a static screen.

Skip Bar screen cleaning rake.

Slide Gate Screenings press by Andritz Sprout-Bauer (Western Hemisphere) and Contra-Shear Engineering, Ltd.

SludgeCleaner Sludge screening and compacting device by Parkson Corp.

Sludgepactor Combination screening, dewatering, and compacting unit by Jones + Attwood.

Sluice gate Manual or power-operated gate used to isolate a channel from flow.

Slush Oil A protective nondrying oil or grease that adheres to steel surfaces, remains soft for prolonged periods, and can be readily removed.

Smooth-tex Rectangular woven wire mesh by Envirex, Inc.

Spirolift Vertical screw conveyor by Spirac Engineering.

Spiropac Compactor by Spirac Engineering.

Spiropress Screenings dewatering device by Spirac Engineering.

Sprout-Bauer Former name of Andritz Sprout-Bauer.

Sta-Sieve Static screen by SWECO, Inc.

Stapling The entanglement of stringy or fibrous debris on a mesh or bar rack.

StarScreen In-channel fine screen by OVRC Environmental (US) and Sernagiotto (Italy).

Stato-Screen Static screen by Vulcan Industries, Inc.

Stellite Family of cobalt-based alloys used as a bushing material by Cabot Corp.

Step Screen In-channel fine screen by Hycor Corp. (USA) and Hydropress Wallender (Sweden).

Stoody Manufacturer of a line of wear-resistant alloy parts. Commonly used term to refer to their centrifugally cast chrome-nickel-boron bushing material.

Stop Log A removable wooden, steel, or concrete bulkhead that fits in vertical grooves in a channel to stop water flow.

Straightline Cable-operated bar screen by FMC Corp.

Suboscreen In-channel rotary fine screen by Andritz Sprout-Bauer (Western Hemisphere) and Contra-Shear Engineering, Ltd.

Supermal High-strength pearlitic malleable iron chain material by Dresser/Jeffery.

Suspended Solids (SS)   Solids captured by filtration through a glass wool mat or a 0.45-micron filter membrane.

Taper-lock Sprocket   Term for sprockets equipped with a split tapered bushing for rigid mounting on a shaft.

Taskmaster   Shedder by Franklin Miller.

Tensile Strength   The maximum tensile load per square unit of cross section that a material is able to withstand.

Tertiary   The use of physical, chemical, or biological means to improve secondary effluent quality.

Thermoplastic   A high polymer that flows and melts when heated. Scrap may be recovered by remelting and reusing.

Thermosetting Plastic   A high polymer that sets into a rigid network. These polymers are not reusable.

Thimble, Chain   Another term used to refer to a chain bushing.

Thru-Clean   Back cleaned, multi-rake bar screen by FMC Corp.

Tines   The tooth or prong of a bar screen cleaning rake.

Totally Enclosed Fan Cooled (TEFC)   Designation for motor enclosure that is not airtight but constructed so as to prevent free exchange of air between the inside and outside of the motor case. Exterior cooling is provided by a fan integral with the machine, but external to the enclosing parts.

Totally Enclosed Nonventilated (TENV)   Designation for motor enclosure that is not airtight, but is constructed so as to prevent free exchange of air between the inside and outside of the motor case.

Toughness   The property of a material that enables it to absorb energy while being stressed above it elastic limit but without being fractured.

Transducer   A device that receives energy from one system and retransmits it, often in another form, to another system.

Tray   The individual screening elements used in a traveling water screen consisting of a wire mesh panel and its structural frame. Also called a "basket."

Tritor   Combination bar screen and grit removal device by FMC Corp.

Trituator   Screenings grinder by Envirex, Inc.

Turbulent Flow   A flow situation in which the fluid moves in a random manner, with a Reynolds number usually greater than 4000.

Turnover   Seasonal (spring and fall) change that occurs in a lake's thermal gradients, resulting in the circulation of biological and chemical materials.

Twilled Weave   Wire mesh woven with each warp and shute wire passing over two and under two of the next adjacent pair of wires.

Ultimate Strength   The stress, calculated on the maximum value of the force and the original area of the cross section, which causes fracture of the material.

VDS Vertical drum screen by Jones + Attwood.

Vee-wire Wedge-shaped wire by Wheelabrator Engineered Systems, Inc./Johnson Screen Products.

Velocity Cap The horizontal cap on vertical, offshore water intake that results in a horizontal inflow, thus reducing fish entrainment.

Verti-Press Solids dewatering press by Dontech, Inc.

Vibrasieve Vibrating fine screen by Andritz Sprout-Bauer.

Vickers Hardness An accurate hardness rating determined by the surface area of the cross section of an indentation made by a pyramidal-shaped diamond pressed into a sample material with a standard force.

Warp The vertical wire in woven wire mesh.

Washpactor Sewage screenings washing and compacting unit by Jones + Atwood.

Waterbox The chamber at the inlet end of a condenser tubesheet.

Wedgewater Sieve Static screen by Hendrick Mfg. Co.

Wedgewire General term to describe trapezoidal or v-shaped wire.

Weft The horizontal wire in woven wire mesh, also called the "shute" wire.

Weir An adjustable baffle over which water flows.

Wiese-Flo In-channel fine screen by Wheelabrator Engineered Systems/Weisemann Screen Products.

Wiesemann Engineering Former name of screening equipment manufacturer acquired by Wheelabrator Engineered Systems/Weisemann Screen Products.

Working Load An allowable recommended tensile load for chains used on conveyors, screens, or other applications of low relative speeds.

Worm A shank having at least one complete thread around the pitch surface.

Worm Gear A gear with teeth cut on an angle to be driven by a worm; used to connect nonparallel, nonintersecting shafts.

Wring-Dry Internally fed rotary screen by Schlueter Co.

Yield That stress in a material at which plastic deformation occurs.

Zebra Mussel Freshwater mollusk that can foul water intake screens and piping by attaching itself to a solid structure, eventually restricting flow.

Z-metal High-strength, pearlitic, malleable iron chain material by Rexnord, Inc.

Bibliography

PUBLICATIONS, BOOKS, AND GENERAL REFERENCES

Agranoff, J. (Editor), *Modern Plastics Encyclopedia*, McGraw-Hill Inc., New York, NY (1982).

Bell, M. C., Mussalli, Y. G., Richards, R. T., Taft, E. P., Wagner, C. H., Watts, F. J. *Design of Water Intake Structures for Fish Protection*, American Society of Civil Engineers, New York, NY (1982).

Belscher, D. "How Dry is Dry Enough?" *Water Environment & Technology*, May 1994.

Borowiec, A. N., Chack, J. J., Lopez, A. R., Melbinger, N. R., *The Bar Screens of New York*, WPCF Operations Forum (1987).

Brady, G. S. and Clauser, H. R., *Materials Handbook*, McGraw-Hill Inc., New York, NY (1978).

Cravens, J. B., Kormanik, R. A., Wittman, J. W., "Innovative Application of Micro-screening in Lagoon Effluent Polishing," presented at *The Sixth Mid-America Conference on Environmental Engineering Design* (1982).

"Evaluating Inlet Screens," *W&WT Magazine*, April, 1993.

Fletcher, R. A., "Risk Analysis for Fish Diversion Experiments: Pumped Intake Systems," *Transactions of the American Fisheries Society* (1985).

Gauthier, H., Jackson, E., LeGrand, A., Simon, R., "Cooling Water Screens and Pumps in French Power Plants," *EPRI Symposium on Condenser Macrofouling Control Technologies: The State of the Art*, Hyannis, MA (1983).

Hazen and Sawyer, Engineers, *Process Design Manual for Suspended Solids Removal*, U.S. Environmental Protection Agency, Washington, D.C. (1975).

Henry, W. E., Wenning, M. E., Van Cott, W., "Solving the Zebra Mussel Dilemma," *Public Works Magazine*, November 1992.

Kemmer, F. N. (Technical Editor), *Nalco Water Handbook*, McGraw-Hill Inc., New York, NY (1979).

Laughlin, J. E. and Roming, W. C., "Design of Rotary Fine Screen Facilities in Wastewater Treatment," *Public Works*, April, 1993.

Majewski, W. and Miller, D. C. (Editors), *Predicting Effects of Power Plant Once-through Cooling on Aquatic Systems*, UNESCO, Paris, France (1979).

McGarry, J. A., Lenhart, C. F., "Screening System Makes Downstream Life Simpler," *WATER/Engineering & Management Magazine*, June (1984).

McNamara, M., Northeast Utilities, et al., "Development of High Performance Traveling Water Screen for Millstone Unit 3 Waer Intake," *EPRI Condenser Technology Conference* (1993).

Mitchell, D. G., "Wastewater Washing Screens out Solids," *Environmental Protection Magazine,* September 1994.

Mussalli, Y. G. and Taft, E. P. "Fish Return Systems," *Proceedings from 25th Annual Hydraulics Division Specialty Conference,* Texas A&M University, College Station, TX (1977).

Mussalli, Y. G., Stone & Webster Engineering Corp., and Hillman, R. E., BATTELLE, *Guidelines on Macrofouling Control Technology,* U.S. Environmental Protection Agency, Washington, D.C. (1987).

Mussalli, Y. G., Stone & Webster, et al., "New Infiltration Intake System for Zebra Mussel Control and Larval Exclusion," *EPRI Condenser Technology Conference* (1993).

Mussalli, Y. G., Stone & Webster, et al., "EPRI Zebra Mussel Monitoring and Control Guide," *EPRI Condenser Technology Conference* (1993).

Nakato, T., Gay, G. E., Kennedy, J. F., *Model Tests of Proposed Intake Structures,* Iowa Institute of Hydraulic Research, University of Iowa (1979).

O'Keefe, W., "Intake Technology Moves Ahead," *Power Magazine,* New York, NY (1978).

Parker, S. P. (Editor-in-Chief), *McGraw-Hill Dictionary of Scientific and Technical Terms,* McGraw-Hill Inc., New York, NY (1984).

PEDco Environmental Inc., *Economic Analysis of Section 316(b) of FWPCA on the Steam Electric Utility Industry,* U.S. Environmental Protection Agency, Washington, D.C. (1977).

POWER Magazine, *Low-cost intake structure saves on upkeep,* Graw-Hill, Inc. (1973).

Public Works Magazine, *A County-Wide Cooperative Water Project,* November, 1991.

Public Works Magazine, *New York City's CSO Strategy,* March, 1993.

Ray, S. S., Snipes, R. L. and Tomljanovich, D. A. *A State-of-the-Art Report on Intake Technologies,* TVA Power Research Staff, Chattanooga, TN (1976).

Richards, R. T., "Fish Screening—The State of the Practical Art," *Proceedings from 25th Annual Hydraulics Division Specialty Conference,* Texas A&M University College Station, TX (1977).

Roberge, John C., "Flow Characteristics of Hydrothane HDPE Trashrack Assemblies" (undated).

Scott, J. B., *Dictionary of Civil Engineering,* Halsted Press, New York, NY (1981).

Smith, R. D., Young, P. J., "EPA Takes Action on CSO," *Water Environment & Technology,* April, 1993.

Strow, D. *Continuous Operation of Traveling Water Screens,* Unpublished, Hales Corners, WI (1987).

US EPA Municipal Operations Branch, *Field Manual for Performance Evaluation and Troubleshooting at Municipal Wastewater Treatment Facilities,* Washington, D.C. (1978).

Updegraff, K. F., "Microscreens Applied to Wastewater Treatment Pond Effluent," presented at *USEPA's Field Evaluation of I/A Technologies* (1986).

Vanagel, M. A., Swapping Clarifiers for Fine Screens, *Waterworld Magazine,* November, 1993.

Wahanik, R. J. "Influence of Ice in the Design of Intakes," *Proceedings from ASCE Specialty Conference on Applied Techniques for Cold Environments,* Fairbanks, AL (1987).

Walsh, K. M. "Screening Your Options," *Water Environment & Technology,* October, 1993.

WEF & ASCE Joint Committee, *Design of Municipal Wastewater Treatment Plants, Manual of Practice 8,* 1992.

WPCF and AWWA Joint Committee, *Wastewater Treatment Plant Design, A Manual of Practice,* Lancaster Press Inc., Lancaster, PA (1977).

"Zooplankton reduced at water treatment works" Reprint from *W&WT,* November, 1990.

CATALOGS, BROCHURES, AND PRODUCT BULLETINS

"Acme Stationary Trash Rake," Acme Engineering Co., Product Bulletin.

"Advanced Water Intake and Screening Equipment," Beaudrey Corp., Product Bulletin.

"Application Report No. 7," Rex Chainbelt Inc., Bulletin 315-256 (1967).

"Aqua Guard Screen," Parkson Corp., Bulletin AG-400 (1982) (1986) (1992).

"Arc Screen Self Cleaning Bar Screen," Infilco Degremont Inc., Bulletin DB 814 (1984).

"Arlat Bar Screens," Arlat Inc., Product Bulletin.

"Arlat Sprial Dewatering Press," Arlat Inc., Product Bulletin.

"Automatic Trashraking, the Cost Effective Solution," Atlas Polar/Hercules Brochure.

"Automatic Water Screens," J. Blakeborough & Sons Ltd., Bulletin 309/B (1976).

"Auto-Rake," Franklin Miller, Product Bulletin.

"BCL Screen," Hycor Corp., Bulletin 878.

"Barminutor," Chicago Pump Products, Bulletin 7641-C.

"Bar Screen Conversion Parts," Flex-A-Gard, Product Bulletin.

"The Berry Trash Rake," L.H. Berry Inc., Product Bulletin.

"E. Beaudrey & CIE," E. Beaudrey & CIE, Product Bulletin.

"Brackett Bosker Weedscreen Cleaner," Brackett Green, Brochure.

"Brackett Green CSO Band Screen," Brackett Green, Brochure Ref. 3007.

"Brackett EVA Back-Raked Screens," Hawker Siddeley Brackett Ltd., Bulletin 2005.

"Brackett Green Sewage Drum Screens," Brackett Green, Brochure Ref. 3005.

"Brackett Green Thru-Flow Band Screen," Brackett Green, Brochure Ref. 3009.

"Brackett Rotating Drum Screens," Hawker Siddeley Brackett Ltd., Product Booklet.

"Brush Raked Curved Fine Screen," William Green, Bulletin WG/S6-6/82 (1982).

"Carpenter Stainless Steels," Carpenter Technology Corp., 2/87/15M Catalog (1987).

"Catenary Bar Screens," Dresser Industries Inc., Bulletin 28186 (1986).

"Centre Infeed Screens," Contra-Shear, Product Bulletin TechArt 723.

"Central Flow Band Screen," Hawker Siddeley Brackett Ltd., Product Bulletin.

"Channel Flow Disintegrator," C&H Brochure R2485.

"Channel Monster II," JWC Environmental, Product Bulletin C350 7/91.

"Clean Shot," CPC Engineering Corp., Product Brochure.

"Climber Screen," Infilco Degremont Inc., Bulletin DB 803 (1982).

"CPC-Clean Shot," Wheelabrator Engrd Systems, Brochure 4133 4/93.

"Cog Rake Bar Screen," FMC Corp., Bulletin 12100, 8-89-5.

"Cont-Flow Bar Screen," John Meunier Inc., Bulletin 10M-5-84 (1984).

"Cont-Flo News Reprint," John Meunier Inc., Bulletin 10-1M-10-81 (1981).

"Contra-Shear Suboscreen," Contra-Shear, Product Brochure.

"CFC Rotary Screen," CPC Engineering Corp., Bulletin CPC 301 (1977).

"Curved Bar Screens," William Green Ltd., Bulletin WG/S5-2/83 (1983).

"Cylindrical Intake Screens," Cook Screen Technologies, Guide CST 2/91/LC.

"Discostrainer," Hycor Corp., Bulletin 857.

"Dontech," Dontech, Inc., Product Catalog.

"Double & Single Flow Bandscreens," William Green Ltd., Product Bulletin.

"Dresser Frontrunner," Dresser Industries, Product Bulletin.

"Drumfilter," Jones + Attwood, Brochure 93.04.

"Drum Screen, Dual-Flow Screen," McGinnes/Royce, Inc., Product Bulletin (1982).

"Dual-Flow TWS Model 57 Service Instructions," FMC Corp., Booklet 724102.

"Dual-Flow Traveling Water Screens," Norair, Brochure.

"Dual-Flow Traveling Water Screens," Envirex, Inc., Bulletin 315-323, 6/86 3M.

"Dynasieve Rotary Screen," Andritz Sprout-Bauer, Bulletin ASB-2/92-G300.

"Electro-Mechanical Rake," Electromecanique Verbandt Andre, Product Bulletin.

"EMU Extractor," EMU-Umwelttechnik, Hof/Saale, Brochure P74/92.

"EMU Lift," EMU-Umwelttechnik, Hof/Saale, Brochure P66/92.

"Envirex Water Screening Equipment," Envirex, Inc., Bulletin 315-334, 2/91-2.5m.

"Epco Communitors," Epco Technical Bulletin 103.

"E.V.A.," Electromecanique Verbandt Andre, Product Bulletin 050/78.92 61.

"Equipment for Water, Sewage and Industrial Waste," Jeffrey Mgf, No. 1052-B (1974).

"FBS," Jones + Attwood, Product Bulletin J.A.S.E.L. 103/2.

"Fine Mesh Screen Retrofits," Beaudrey Corp., Product Bulletin.

"Fine Screens," Vulcan Industries, Inc., Product Brochure.

"Flo-Screen," Enviro-Care Co., Product Bulletin.

"Flowminutor," Enviro-Care Co., Product Bulletin 3K84 (1984).

"FMC Traveling Water Screens," FMC Corp., Product Bulletin.

"Frontrunner," Dresser Industries, Catalog #28588, 3/88.

"The FS-304 Self-Cleaning Filter Screen," Wiesemann Engr, Product Bulletin.

"FSM-Filterscreen," FSM-Frankenburger, Product Brochure.

"Griger," Hellmut Geiger GmbH & Co., Product Brochure 11-26 176/9 (1987).

"Getting solids Out," Bauer Combustion Engineering Inc., Product Bulletin.

"Grind Hog Comminutors," G.E.T. Industries Product Bulletin.

"Grind Hog Auto-Coupling Slide Rail System," G.E.T. Industries Product Bulletin.

"Water Screening, Plant Design & Layout," E. Beaudrey & CIE, Product Bulletin.

"The Grabber Automatic Bar Screen," Hycor Corp., Bulletin 872.

"Helix Screen and Compactor," Hercules Systems, Ltd., Product Bulletin.

"Here Comes the Brownie Buster," Enviro-Care Product Brochure.

"The Hendrick Wedgewater Sieve," Hendrick Mfg Co., Bulletin HFS 10815M.

"Hercules Hydrorake System," Atlas Polar Company Ltd., Product Bulletin.

"Hercules Systems Screens for Solids Separation," Hercules Systems, Ltd., Product Bulletin.

"Hercules Washing/Dewatering Compactor," Hercules Systems, Ltd., Product Bulletin.

"Hercules 5mm Fine Screen," Hercules Systems, Ltd., Product Bulletin.

"Hydrasieve Screens," Andritz Sprout-Bauer, Bulletin ASB-2/92-PEG-440.

"JWC Environmental," JWC Environmental, Product Bulletin C201 7/91.

"HTG Hydroscreen," Hoelschertechnic-Gorator GmbH Product Brochure.

"HTG Rotoshear," Hoelschertechnic-Gorator GmbH Product Brochure.

"Hydroscreen," Hycor Corp., Product Bulletin.

"Hypress," Hycor Corp., Product Bulletin.

"Hyveyor," Hycor Corp., Bulletin 874.

"Introducing Hawker Siddeley Brackett," Hawker Siddeley Brackett, Publication 2000.

"Involute Panel Bandscreens," William Green Ltd., Bulletin WG/S1-2/83.

"IPEC IFS Series," IPEC Industries, Product Bulletin.

"IPEC RSS Series," IPEC Industries, Product Bulletin.

"Johnson Surface Water Intake Screens," UOP Inc., Product Bulletin 201081 (1981) & (1987).

"The Kason Cross-Flo Sieve," Kason Corp., Bulletin CF-79.

"Landustrie Screen Cleaners,," Landustrie, Brochure 02-93-600 ENG.

"Level Measuring Systems," Milltronics, Bulletin 12/83-1M (1983).

"Lift Type Mechanically Cleaned Bar Screens," Envirex Inc., Bulletin 315-22 9/85.

"Link-Belt Four-Rope Bar Screens," FMC Corp., Booklet 12240.80.

"Link-Belt Screens," FMC Corp., Bulletin 12000.

"Link-Belt Traveling Water Screens," FMC Corp., Bulletin 41006.

"Lo Flo Radial Bar Screen," William Green Ltd., Product Bulletin.

"Mechanically Raked Bar Screens," William Green Ltd., Bulletin WG/S4-2/83 (1983).

"Mechanical Rake Screen," Ashbrook-Simon-Hartley, Product Bulletin.

"Mensch Bar Screen," Vulcan Industries, Brochure B94-120-M.

"Micrasieve Screens," Andritz Sprout-Bauer, Bulletin ASB-2/92-PEG-440.

"Micromesh Strainer," Lakeside Equipment Corp., Bulletin 230 (1977).

"MICRO-Pi," Contra-Shear, Product Bulletin.

"Microscreens," Envirex Inc., Bulletin 315.31 3M-11/84 (1981).

"Microscreens," Hawker Siddeley Brackett Ltd., Ref. No. 2012.

"Microstraining," Weir Pumps, Ltd. Publication #AP47A (1987), #AP51, #AP52, #AP53.

"Microstraining and its Applications," Weir Pumps, Ltd. Publication AP1/R3 (1986).

"Microstraining Anti-Pollution Systems," Weir Pumps, Ltd. Publication AP47A.

"The Milliscreen Treatment Plant," Contra-Shear Developments Ltd., Publication.

"Mobius Screening System," Pro-Ent, Inc., Product Bulletin.

"Model 572 Level Measurement System," Ametek Inc., Bulletin DB-572 (1987).

"Muffin Monster," Disposable Waste Systems Inc., Bulletin F2-MM1180.

"Muffin Monster," JWC Environmental, Product Bulletin C401 9/91.

"Net-Waste," Diemme USA, Product Brochure.

"Non-Metallic Baskets," Screening Systems Int'l, Product Bulletin.

"No-Well Traveling Water Screen," FMC Corp., Bulletin 814101 (1978).

"No-Well Traveling Water Screens Service Instructions," FMC Corp., Booklet 624101.

"Original Chicago Pump Comminutor," Yeomans Chicago Corp. Bulletin 7610-D.

"Ossberger Trash Rack Cleaning Systems," Ossberger Turbines Inc., Product Bulletin.

"Parkwood Inlet Screens," The Longwood Engineering Co., Publication LE/3/86 (1986).

"Passavant Cable Operated Bar Screen," Passavant-Werke AG & Co., Product Bulletin.

"Passavant Cooling Water Screening," Passavant-Werke AG & Co., Product Bulletin.

"Passavant Micro-Sieve," Passavant Corp., Bulletin 1661 (1975).

"Passavant Posirake Model 1230," Passavant Corp., Bulletin 1231 (1975).

"Passavant Travelling Band Screen," Passavant-Werke AG & Co., Product Booklet.

"Permutit Micro-Matic Rotating Drum Strainers," Permutit, No 6210-982-5M (1982).

"Polyurethane Bar Screen Rakes," Flex-A-Gard, Product Bulletin.

"Posirake Bar Screen," Passavant Corp., Product Booklet (1986).

"Powermatic Screen," William Green Ltd., Bulletin WG/S3-2/83 (1983).

"The Power of the Roller Chain!" Vulcan Industries Inc., Product Bulletin.

"Powrclean Bar Screen," Peabody Welles, Bulletin PW 0100-1 (1978).

"Preliminary Treatment," Biwater Sewage Treatment, Product Bulletin.

"Pressveyor," Hycor Corp., Product Bulletin.

"Purac Screenings Press," Purac, Bulletin SF-120/0591.

"Purpose and Application of S&L Bar Screens," Smith & Loveless Engr Data (1977).

"Radial Bar Screen," William Green Ltd., Bulletin WG/S7-3/83 (1983).

"Ranney Water Collector Systems," Hydro Group, Inc., Bulletin 1M-11-83 (1983).

"729 PS Reinforced Polyurethane Chain," Flex-A-Gard, Product Bulletin.

"Replacement Parts and Service," Power Plant Service Supply, Parts List (1987).

"Rex Heavy-Duty Bar Screens," Rex Chain Belt Co., Bulletin 315-22 (1961).

"Rex Mechanically Cleaned Bar Screens," Envirex Inc., Bulletin 315-21 & R1 4/91.

"Rex Power Transmission Components," Rexnord, Inc., R85 Catalog (1984).

"Rex Traveling Water Screens," Rex Chain Belt Co., Catalog #147 (1927), #187 (1929), #336 (1938).

"Rex Water Intake Screens," Envirex Inc., Bulletin 315-331,1/90-3M & 315-331R1,8/92.

"Rex Water Screening Equipment," Envirex Inc., Bulletin 315-321.

"Robo Bar Screen," Vulcan Industries, Brochure B94-120-R.

"Rotafine," Jones + Attwood, Product Bulletin J.A.S.E.L. 116/2.

"Rotamat Screens," Lakeside Equipment Co., Product Bulletin.

"Rotamat Wash Press," Lakeside Equipment Corp., Product Bulletin.

"Rotarc Brush Screens," John Meunier Inc., Bulletin 7-2M-10-81 (1981).

"Rotarc Mini Screens," John Meunier Inc., Product Bulletin.

"Rotary Screens and Microstrainers," William Green, Ltd., Bulletin WG/S2-2/83 (1983).

"Rotasieve," Jones + Attwood, Product Bulletin J.A.S.E.L. 127/1.

"RotoClean," Parkson Corp, Bulletin RC.

"Roto-Guard," Parkson Corp, Bulletin RG701.

"The Rotopac Screw Compactor," John Meunier Inc., Bulletin U-M1-3m-6-83 (1983).

"RotoPress," Parkson Corp, Bulletin RP1401 492.

"Rotoshear," Hycor Corp., Bulletin 105-R (1983).

"Rotostep," Hycor Corp., Bulletin 865.

"Rotostrainer," Hycor Corp., Bulletin RS 1101 483 (1977).

"RR Continuous Screw Dewatering Presses," AMETEK, Bulletin RR-IND (1) (1986).

"Safgard Rotary Drum Strainer," Schlueter Co., Brochure.

"Schrage Gegenstromrechen," Friedrich Schrage, Product Bulletin.

"Schreiber Frontloader Screen," Schreiber Corp., Product Bulletin.

"Screening Equipment," UBE Industries Ltd., Product Bulletin.

"Screening Equipment," FMC Corp., Book 2587A (1966).

"Scrrenings Systems Curved Bar Screen," Kruger, Product Bulletin A0621-1000-1992.

"Screening Technology by Contra-Shear," Contra-Shear Ltd., TechArt 213 L.

"Screezer," Jones + Attwood, Product Bulletin J.A.S.E.L. 1101/3.

"Self-Cleaning Trashrack," The Duperon Corp., Product Bulletin.

"1500 Series Screens," Contra-Shear Ltd., Product Bulletin TechArt 814/85.8.

"Service Instructions for Link-Belt TWS," FMC Corp., Booklet 2690 (1973).

"Semi-Rotary Bar Screen," William Green Ltd., Bulletin WG/S9-3/83 (1983).

"Silmattan Z-100," Zickert Products, Product Bulletin 05/91.

"Single & Double Flow Band Screens," Ledward & Beckett Ltd., Product Bulletin.

"Single Tank Level Measuring System," Inventron McGraw-Edison Co., Section 9100.

"SludgeCleaner," Parkson Corp, Bulletin SP1311 292.

"Sludgepactor," Jones + Attwood, Product Bulletin J.A.S.E.L. 123/1.

"Smalley Trash Rakes," Smalley Excavators Ltd., Product Bulletin.

"Spiropress," Sprirac Engineering AB, Product Bulletin 9305.

"Spray Nozzles," FMC Corp., Folder 32301 (1973).

"Spray Nozzle User's Manual," Spraco Inc., Catalog 8507 (1985).

"Stormwater Overflow Screens," John Meunier Inc., Product Bulletin.

"Straightline Bar Screen," FMC Corp., Folder 2945 (1964).

"The Tallest Cont-Flo Bar Screen," John Neumier Inc., Product Bulletin (1982).

"Taskmaster," Franklin Miller Inc., Product Bulletin.

"Test Results Using Contra-Shear Screens," Contra-Shear Ltd, Publication (1983).

"Tetko Fabrication Services Guide," Tetko Inc.

"There's gold in them there screens," Hycor Corp., Product Bulletin.

"Traveling Screens to Protect Fish," Envirex, Inc., Bulletin 316-300 (1973).

"Traveling Water Screens," FMC Corp., Catalog 2652, #51009 (1975), #710101 (1987).

"Traveling Water Screens Fish Protection System," FMC Corp., Bulletin 41006 (1974).

"Traveling Water Screen Model 45A Thru-Flow," FMC Corp., Bulletin 41007 (1974).

"Traveling Screen Control," Drexelbrook Engr Co., Bulletin 303-600-A (1987).

"Typical Biological Waste Treatment System," Smith & Loveless Inc., Bulletin S-121.

"The VDS," Jones + Attwood, Product Bulletin J.A.S.E.L. 100/3.

"Washpactor," Jones + Attwood, Product Bulletin.

"Wastewater Disintegrating and Screening," Franklin Miller Inc., Product Bulletin.

"Waste Water Treatment Equipment," Fairfleld Service Co., Product Bulletin.

"Wastewater Disentegrating and Screening," Franklin Miller Inc., Booklet.

"Water," Davy Bamag GmbH, Product Bulletin.

"Water Screening Systems Delivery Program," Esmil Hubert BV.

"Water Supply, Treatment," VIZGEP, Product Brochure.

"Water Treatment Equipment," Smith & Loveless Inc., Bulletin 1627MG.

"Z-1100 Screenmat," Zickert Products, Bulletin GBG 04.92.

Index

Milton Keynes UK
Ingram Content Group UK Ltd.
UKHW040447071024
449327UK00020B/1051